A Field Guide to the

SOUTHEAST COAST & GULF OF MEXICO

A Field Guide to the

SOUTHEAST COAST & GULF OF MEXICO

COASTAL HABITATS, SEABIRDS, MARINE MAMMALS, FISH, & OTHER WILDLIFE

NOBLE S. PROCTOR

PATRICK J. LYNCH

Illustrated by
PATRICK J. LYNCH

Yale

UNIVERSITY

PRESS

Yale University Press books may be purchased in quantity for educational, business, or
promotional use. For information, please e-mail sales.press@yale.edu (US office) or sales@
yaleup.co.uk (UK office).

Designed by Patrick J. Lynch.
Set in Adobe Utopia and Adobe Univers type.
Printed in China.

ISBN 978-0-300-11328-0
Library of Congress Control Number: 2011931241

This paper meets the requirements of ANSI/NISO Z39.48-1992 (Permanence of Paper).

10 9 8 7 6 5 4 3 2 1

To Braxton and Alexis:
youthful interest and excitement personified.
May you enjoy all aspects of
natural history throughout your life.

To Susan, Devorah, Alex, and Tyler
and to Noble: teacher, mentor, and friend.

CONTENTS

ix   *Preface*

xi   *Introduction*

xxii   *Regional maps*

xxvi   *Topography of marine animals*

2   Marine and coastal plants and habitats

30   Invertebrates

62   Sharks

82   Rays

88   Fish

182   Sea turtles

188   Crocodile and alligator

190   Marine and coastal birds

320   Baleen whales

334   Toothed whales and dolphins

366   Seals and manatee

373   *Index*

Great Blue Heron, Everglades National Park, P. Lynch

PREFACE

This book has been several years in the making, but in a sense we've been preparing it for forty years and more—on countless pelagic birding and whale-watching trips and on coastal birding excursions from Corolla, North Carolina, down to the Dry Tortugas and many spots along the Gulf Coast. We couldn't possibly list the captains, crews, and birding companions of every trip here, but we thank them all for the opportunity to view and enjoy the offshore environment and animals of the southeastern and Gulf coasts.

We also thank the fishing parties of day boats, dock fishermen, and crew members of fishing boats for patiently allowing us to look at, poke and prod, photograph, and discuss the many interesting fish they brought ashore or caught and released. And we are very grateful to the hundreds of fellow birders and whale watchers we have shared a deck with, and for the privilege of sharing their knowledge, enthusiasm, and deep concern for the fate of our fragile coastal environments.

Many individuals have generously shared their data and knowledge of specific species and have provided support for this project. During various pelagic trips Lyn Atherton, Davis Finch, Frank Gallo, Frank Gardiner, Dr. Charles "Stormy" Mayo, Dr. Roger Payne, Diana Payne, Fred Sibley, Don Sinotti, Dr. Jeffrey Spendelow, James Stone, and Dr. Frank Trainor have shared their time and expertise with us. Sincere thanks from Pat to Captain Bob Bates, Captain Tom, Mike, Andy, and the staff at Horizon Divers in Key Largo for the many great dive trips to Pennekamp area coral reefs.

Those who have lent their support, expertise, photographs, and comments as the book took shape include Margaret Ardwin, David Bolinsky, William Burt, Daniel Cinotti, Frank Gallo, Sarah Horton, Carl Jaffe, Janet Jeddrey, Rick Leone, Devorah Lynch, Kimberly Pasko, Wayne Petersen, Roger Tory Peterson, Fred Richards, Sally Richards, Fred Sibley, Dr. Jeffrey Spendelow, and Alex and Tyler Wack. We greatly appreciate everything these friends have contributed to this field guide.

We extend a special thank you to Jean Thomson Black, executive editor for life sciences at Yale University Press, for her faith in us over the years and for being our champion at the Press to see that this book came to fruition.

We also thank our manuscript editor, Laura Jones Dooley, for her thorough dedication to editorial quality and for always making us seem much more articulate than we really are.

PREFACE

And, of course, our special thanks go to our wives and families:

For Noble—To Carolyn, who has the patience of a saint, and to Adam and Eric, who continue to show unequaled support in all I do. Without their love and interest, my accomplishments would be empty.

For Pat—To dearest Zhu, and to Devorah, Alex, and Tyler, who make it all worthwhile, and to Jagger, Navi, Tex, and Kitty, our wildlife in residence.

NOBLE S. PROCTOR
Branford, Connecticut

PATRICK J. LYNCH
North Haven, Connecticut
patricklynch.net
coastfieldguides.com

INTRODUCTION

This field guide covers the major marine and coastal life of the southeastern Atlantic Coast from Cape Hatteras, North Carolina, to the Florida Keys and Gulf Coast of the United States. In our previous book, *A Field Guide to North Atlantic Wildlife,* we ended southern coverage at Cape Hatteras because the cape is a major biological boundary between the cold, northwestern Atlantic seas, dominated by the Labrador current, and the warm, tropical influence of the Gulf Stream as it passes by the cape. Biologically and geologically the southeastern Atlantic and Gulf Coasts form a continuous floral and faunal coastal region from the Carolinas to Texas.

Within the limits of a guide meant for field use we cannot cover every marine and coastal species of plant and animal. This book is designed as a general reference for birders, divers, sportfishing enthusiasts, boaters, kayakers, beach hikers, and anyone else curious about the natural world and its plants and animals. Birds especially are a highly visible, diverse, and accessible form of wildlife that cannot be covered completely in a general guide, so we urge birders to carry along their favorite bird guide as a supplement to this book. Even if your passion is for a particular group of animals, we ask you to think in broad environmental terms, to look holistically at the coastal and offshore waters, and to see and appreciate the great cycles of migration and weather, human and natural activity, seasons and tides, that make the shoreline endlessly fascinating.

The Gulf Stream

The Gulf Stream, along with its antecedent currents in the Caribbean Sea and the Gulf of Mexico, is the dominant geophysical feature of the southeastern US coast. One of the fastest, largest, and warmest of all ocean currents, the Gulf Stream carries tropical heat from the Florida Keys all the way to the Hebrides Islands

The southeastern coast is beautiful but delicate. Nag's Head Woods Ecological Preserve, along Roanoke Sound, North Carolina.

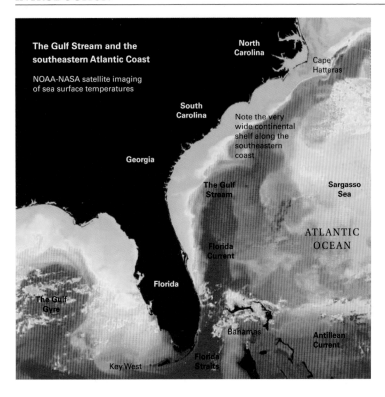

The Gulf Stream and the southeastern Atlantic Coast

NOAA-NASA satellite imaging of sea surface temperatures

North Carolina

Cape Hatteras

South Carolina

Note the very wide continental shelf along the southeastern coast

Georgia

The Gulf Stream

Sargasso Sea

ATLANTIC OCEAN

Florida Current

Florida

The Gulf Gyre

Bahamas

Antillean Current

Key West

Florida Straits

off Scotland. Off the Florida Keys, the stream flows at 35.3 million cubic feet (1 million cubic meters) of water per second, and by the time it veers eastward around the Grand Banks of Newfoundland, its core is a hundred miles wide and moves 5.3 billion cubic feet (150 million cubic meters) of water per second. The Gulf Stream carries more water than the combined flow of all of the rivers that empty into the Atlantic Ocean. Although Benjamin Franklin is famous for having named and publicized the existence of the Gulf Stream in the 1770s, the current was known to Spanish mariners at least as early as Ponce de León, who noted it in 1513, and the Gulf Stream became the major eastward route for galleons returning to Spain from the Americas.

The Gulf Stream is intimately related to other currents in the Atlantic Ocean, Caribbean Sea, and Gulf of Mexico and is not simply the drainage from the Gulf. It is best thought of as the northern continuation of the North and South Equatorial Currents that feed into the Caribbean Sea and the westernmost arc of the great clockwise circular current of the North Atlantic basin, the North Atlantic Gyre.

Migration
Every spring and fall the US coastline plays host to millions of migrants. Many

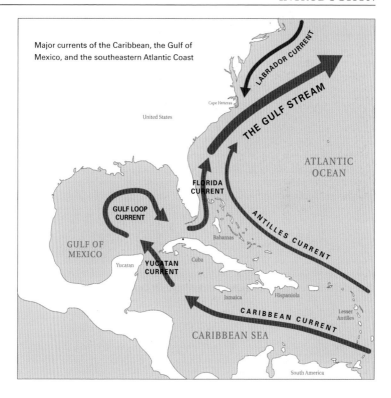

Major currents of the Caribbean, the Gulf of Mexico, and the southeastern Atlantic Coast

species of birds, fish, marine mammals, and even insects like dragonflies and the Monarch Butterfly undertake long seasonal migrations along the Atlantic and Gulf Coasts. Land birds arriving from the south seek shelter and food along our shorelines, and areas such as High Island, Texas, have become vital stops for the survival of migrating birds.

On our beaches, bays, and coastal marsh edges, thousands of shorebirds and ducks stop to rest and feed on their journey northward, many traveling as far as the high Arctic to nest. Shorebirds blanket the coasts as they rest on their long journey from the south. Many feed on insects in the wrack line, as well as on eggs or on crabs and other invertebrates. Some species, such as the Red Knot, time their migrations to coincide with the spring egg-laying season of the Horseshoe Crab. Without these seasonal food sources, bird populations quickly dwindle to critically low numbers.

For migrant birds to survive, our coastal environments must remain intact. It is of paramount importance that we ensure the availability of feeding and resting sites for migrant populations so that they may reach their breeding and wintering grounds. Land development, oil disasters, and chemical pollution through the improper disposal of waste products all contribute to the loss of shoreline habitats.

The southeastern Atlantic Coast

The exposed ocean shorelines of the southeastern coastline consist almost entirely of sandy beaches, either on the mainland or on barrier islands just off the coast. The many large rivers that flow into the coastal waters of the Carolinas and Georgia also support huge salt marshes along the river mouths and behind the barrier islands. The relatively gentle slope of the continental shelf, combined with the huge amount of sediment delivered by the southeastern rivers, creates a shallow, turbid, nutrient-rich marine environment that supports a vast range of coastal and marine wildlife. Although it flows as much as seventy miles off the shores of Georgia and South Carolina, the Gulf Stream forms the eastern ecological boundary of the southeastern coastal environment, bringing warmth and extending the northward range of tropical species.

Most of the species described in this field guide inhabit the relatively shallow waters of the continental shelf or inshore coastal areas. Off the Carolinas and Georgia the continental shelf is so shallow that the 100-fathom (600 ft.) line of the shelf edge can be as much as fifty miles offshore. This is why the great whales and such truly marine birds as shearwaters, though common in the southeastern Atlantic, are seldom seen within miles of land. These species prefer the deep oceanic waters at the edge of the continental shelf, although Humpback, Right, and Fin Whales also hunt the shallower fish-rich areas of offshore banks.

Barrier islands and capes

As an environment dominated by the action of waves and the movement of sand, the southeastern coast is highly changeable, and major variations in coastal features are noticeable even within a human lifespan. Sandy barrier islands are usually formed by longshore currents that carry sand along the coastline and build up sand deposits in shallow areas or, in Florida, on the remains of ancient

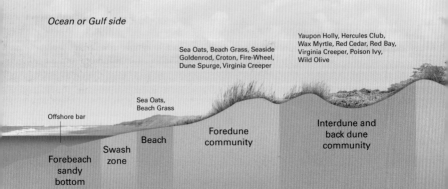

Ocean or Gulf side

Yaupon Holly, Hercules Club, Wax Myrtle, Red Cedar, Red Bay, Virginia Creeper, Poison Ivy, Wild Olive

Sea Oats, Beach Grass, Seaside Goldenrod, Croton, Fire-Wheel, Dune Spurge, Virginia Creeper

Sea Oats, Beach Grass

Offshore bar

Foredune community

Interdune and back dune community

Beach

Swash zone

Forebeach sandy bottom

coral reefs that run parallel to the coast. Longshore drifts of sand in water and wind-driven sand gradually form deposits that stay above the high tide line, and pioneer plants take root. Plants are crucial to the formation of barrier islands: their leaves trap wind-borne sand and help build up the island. Tidal flow also acts to build barrier islands. Swift offshore currents slow as they flow around barrier islands, dropping loads of sand and sediment onto the sheltered sides of islands.

As sand blows inland from the beach, it forms dune lines that run parallel to the coast. The seaward foredunes shelter plant life farther from the beach, decreasing the plants' exposure to salt spray and creating protected pockets where less salt-tolerant floral communities can form. The salt-spray zone near the exposed ocean beach is lethal to all but specially adapted plants. Near the beach a fine, nearly invisible layer of salt coats every leaf surface. Salt crystals also act as lenses that focus the sun's rays and burn leaf surfaces.

Behind the first dunes, plant communities form a predictable sequence that is visible on barrier islands from Cape Hatteras, North Carolina, all the way to Padre Island, Texas. Beach pioneer species like Sea Oats give way to creeping vine species and low shrubs on the interior dunes, and those dunes in turn shelter maritime forests on the larger barrier islands. In the protected areas on the landward sides of barrier islands, salt marshes form on the fine silt deposited by slower-moving waters, completing an environmental sequence from ocean side to bay side that can take hundreds of years to create from bare sand.

The southeastern coast is also characterized by a series of coastal capes formed by the same longshore drift, tidal action, and plant communities as the smaller barrier islands. Cape Hatteras, Cape Lookout, Cape Fear, and Cape Canaveral are the largest capes along the southeastern Atlantic Coast, and smaller sandy capes such as Cape San Blas also form along the Gulf Coast. These capes often

Loblolly and Longleaf Pines, Dogwood, Live Oak, Yaupon Holly, Red Cedar

Bay or sound side of the island

Saltmeadow Cordgrass, Black Needlerush, Sea Ox-Eye, Glasswort

Smooth Cordgrass, Black Needlerush

Maritime forest

Upper salt marsh

Lower salt marsh

Brackish water

mark biological boundaries. For example, Florida's Cape Canaveral is roughly the southern boundary of the warm temperate coastal ecology of the Southeast and the northern edge of the West Indian faunal province, where mangroves and other tropical species begin to dominate the coast and coastal waters.

The Gulf of Mexico and Gulf Coast

The geology and biology of the Gulf Coast is very similar to the southeastern Atlantic Coast: a gently sloping coastal plain leading out to a very wide, shallow, muddy continental shelf that in many places is less than twenty feet deep even many miles from shore. Physically the Gulf Coast is split between the sandy coastlines of Florida and Texas and the muddier, marsh-dominated coasts around the mouths of the Mississippi and Alabama Rivers, Mobile Bay, and the Atchafalaya River complex that feeds Atchafalaya Bay. These rivers pour millions of tons of sediment into the Gulf, two million tons per day from the Mississippi River alone. From the Florida border in the east to the Texas border in the west, the central Gulf Coast is more a scarcely defined mix of marsh islands and river delta channels than a hard coastline, but this rich mix of sediments, estuaries, and shallow, warm salt water make the Gulf Coast one of the most productive fisheries and important wildlife areas in the world.

The Gulf of Mexico is a deep basin that has been geologically stable for millions of years. The Gulf waters are fed in the south by the Yucatán Current flowing through the Yucatán Channel. The Yucatán Current is the primary motive force behind the Gulf Gyre, a roughly circular, clockwise flow of extremely warm water that has become notorious recently as the heat engine that adds tremendous destructive power to the great hurricanes that cross the Gulf of Mexico every decade or so. Three of the most damaging and lethal storms in US history, the Galveston Hurricane of 1900, Hurricane Camille in 1969, and Hurricane Katrina in 2005, all

A NASA GEOS-12 infrared satellite image of Hurricane Katrina just before it made landfall in southeastern Louisiana on August 29, 2005. NASA Goddard Space Flight Center.

strengthened into Category 5 monsters as they passed over the hot Gulf Gyre on their way toward the US coastline. Another recent Category 5 storm, Hurricane Mitch, gained both heat energy and tremendous amounts of moisture as it crossed the southern Gulf and Yucatán Channel, and the rains and high winds it brought to Central America killed 11,000 people in 1998.

The Deepwater Horizon disaster

On April 20, 2010, a catastrophic explosion and fire struck BP's Deepwater Horizon oil drilling platform, and it sank two days later. The resulting blowout of the Macondo well lasted 88 days before engineers finally capped it on July 15, 2010. During the Macondo well blowout, an estimated 12,000–19,000 barrels of oil entered the Gulf of Mexico each day. The total spill from the disaster was just under five million barrels of oil, making it the largest marine oil spill in history. In a report to President Barack Obama, an independent commission of experts concluded that "regulators failed to keep pace with the industrial expansion and new technology—often because of industry's resistance to more effective oversight. The result was a serious, and ultimately inexcusable, shortfall in supervision of offshore drilling that played out in the Macondo well blowout and the catastrophic oil spill that followed."

The effects of the Macondo well blowout will take decades to determine. Not all the news is bad: much of the oil spilled was composed of volatile elements that evaporated at the surface, were burned off, or were quickly broken down by sunlight, wave action, and marine bacteria and algae that can feed on hydrocarbons. However, millions of tons of heavier oil components remain in Gulf waters, and these biologically active hydrocarbons will be in the ecosystem for many years to come, with unknown long-term consequences.

US Coast Guard photograph, April 21, 2010

INTRODUCTION

Climate change

No one who pays any attention to the natural environment could fail to notice the effects of global warming on the distribution of coastal animals. Formerly "southern" species such as pelicans, oystercatchers, avocets, skimmers, and others are year by year extending their summer ranges north along the coast. Even delicate and specialized species like the West Indian Manatee are now regularly seen along the mid-Atlantic Coast, something that would have been unthinkably odd just twenty years ago.

Brown Pelicans along the Louisiana coast, with an oiled bird on the left. Photo courtesy William Burt © 2010.

Climate change, along with habitat destruction and changes in prey populations, has also been a factor in more subtle changes in animal behavior and health. Seal species seem to be shifting southward every year, and even a North Atlantic species like the Gray Seal has been appearing rarely but regularly as far south as Cape Hatteras. Scientists are suspicious that a combination of warmer waters and increased coastal pollution is responsible for our frequent "red tide" algal blooms along the coasts. Habitat changes may also play a part in various disease epidemics that have affected coastal dolphin, seal, lobster, and oyster populations recently.

Overfishing and habitat destruction

If you are reading this book, you are probably at least a casual visitor to ocean beaches, coastal birding environments, dive spots, or fishing areas. Superficially the ocean and coast look the same as they always have: beautiful beaches and coastal islands edging a seemingly unlimited ocean beyond. But within the lifetime of today's adults our coasts have been the scene of an unprecedented massacre and die-off of wild ocean species. As you flip through the pages of this book, notice that virtually every marine fish and mammal longer than three feet is titled in red, because the International Union for Conservation of Nature (IUCN) Red List has identified the species as threatened or endangered.

Many marine fish species are faced with critical drops in population, some to the point of possible extinction. The average body size of most commercially caught fish species has diminished sharply in the past two decades, and in some instances entire regions of the coast have lost what were once productive fisheries. Our demand for fresh seafood has far outstripped the capacity of even the richest fishing grounds.

The Bluefin Tuna is a good example of a species devastated by recent human culinary fads—in this case a demand for luxurious fatty sushi. Bluefin Tuna was never a major element of traditional Japanese sushi culture until factory fishing and air transport made it more available in the 1950s and 1960s. When an adult Bluefin carcass caught in US or European waters can fetch over $100,000

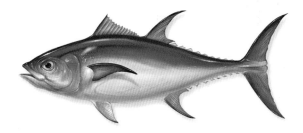

in a Japanese fish market, the prospects for the species are grim. Fifty years of increasing demand has placed this once-abundant ocean predator on the Red List in all its habitats. The young of prized fish like the Bluefin are totally dependent on our coastlines and marshes as nursery systems as they develop. It is more important than ever to ensure that these coastal habitats receive protection from overfishing and habitat destruction along our coasts if wild-caught species of fish and shrimp are to survive. Ironically, we've only recently discovered how critical the Gulf of Mexico is for Bluefin spawning, just in time to see that very area devastated by the Deepwater Horizon disaster.

Habitat destruction is a major but less visible killer of coastal wildlife. Aside from the increasingly frequent massive fish die-offs and red tides of summer, you won't notice the effects of coastal development in dramatic ways. If we take away the environments where coastal animals and marine species live, they simply can't reproduce and will gradually disappear. The Piping Plover and Roseate Tern are examples of species that have become endangered because we've taken away the beach habitats they need for nesting. No one deliberately kills plovers and terns, but year by year the birds quietly fade away as new marinas and shoreline condominiums supplant their homes.

However, not all human effects on the oceans are destructive. One recent trend that has benefited sport divers, fishing enthusiasts, and many marine wildlife species is the creation of artificial reefs. Old naval and commercial ships, obsolete railroad cars and aircraft, and even purpose-built hollow concrete "reef balls"

P. LYNCH

have been sunk off our coasts to create new reef environments. These structures are quickly adopted by bottom invertebrates and fish species and have a tremendous positive effect on the number and variety of marine species nearby. Even oil platforms, bridges, and piers can enhance the underwater environment, providing shelter to species that would otherwise be absent from the area owing to the lack of natural cover.

Red List species
The common names of endangered or threatened marine species listed in this guide are printed in red. For more information on each species, see the IUCN Red List of Endangered Species Web site at www.iucnredlist.org.

About the range maps
The maps in this field guide are general approximations of the seasonal ranges of each species. In many marine species the exact extent of their seasonal and geographic ranges is not well documented, and fish, whales, and seabirds are highly mobile creatures that move freely throughout the Atlantic Ocean and Gulf of Mexico, following winds and weather, ocean currents, and seasonal movements of their prey species.

Cattle Egrets, Bar B Ranch Preserve, Winterhaven, Florida. Sometimes the wildlife win one: Cattle Egrets are thought to have migrated on their own from western Africa to South America in the late 1800s, and by the late 1950s they had spread through Central America and become established in the southern United States. Photo by Noble Proctor.

SOUTHEASTERN COAST AND GULF OF MEXICO

The southeastern US and Gulf Coasts make up some of the richest wildlife areas in the world. From the sandy shores of North Carolina's Outer Banks to the barrier islands of Texas, this vast area encompasses temperate coastal forests, broad, shallow continental shelves, semitropical beaches, and tropical mangrove forests, as well as some of the most productive fishing grounds in the United States.

The nearly landlocked Gulf of Mexico is the ninth largest body of water in the world. The wide continental shelves of the Gulf of Mexico have been particularly exploited for their oil deposits, and huge numbers of oil platforms dot the seas just south of the Louisiana coast. The Gulf's warm, semitropical waters energize the hurricanes that pass over it nearly every year.

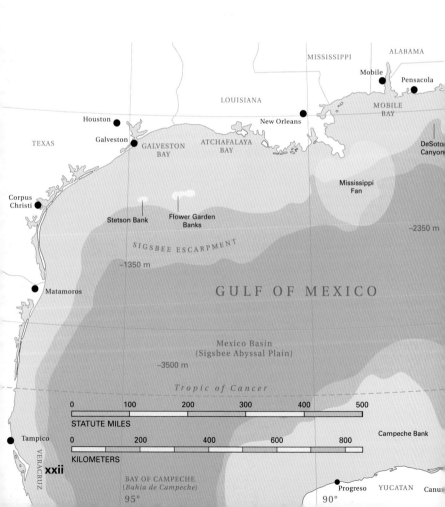

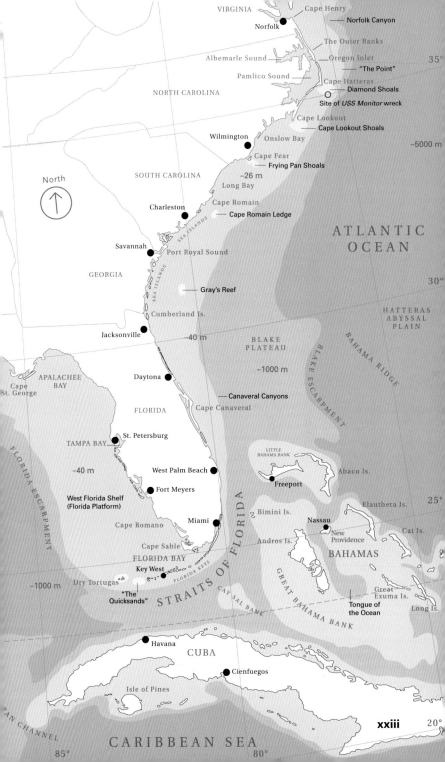

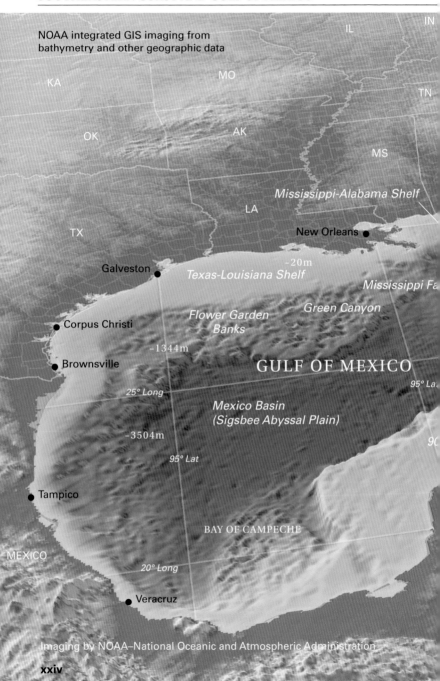

NOAA integrated GIS imaging from
bathymetry and other geographic data

IL

IN

KA

MO

TN

OK

AK

MS

Mississippi-Alabama Shelf

LA

New Orleans ●

TX

−20m

Galveston ●

Texas-Louisiana Shelf

Mississippi Fa

Green Canyon

*Flower Garden
Banks*

Corpus Christi ●

−1344m

GULF OF MEXICO

Brownsville ●

95° La.

25° Long

*Mexico Basin
(Sigsbee Abyssal Plain)*

−3504m

9(

95° Lat

Tampico ●

BAY OF CAMPECHE

MEXICO

20° Long

Veracruz ●

Imaging by NOAA–National Oceanic and Atmospheric Administration

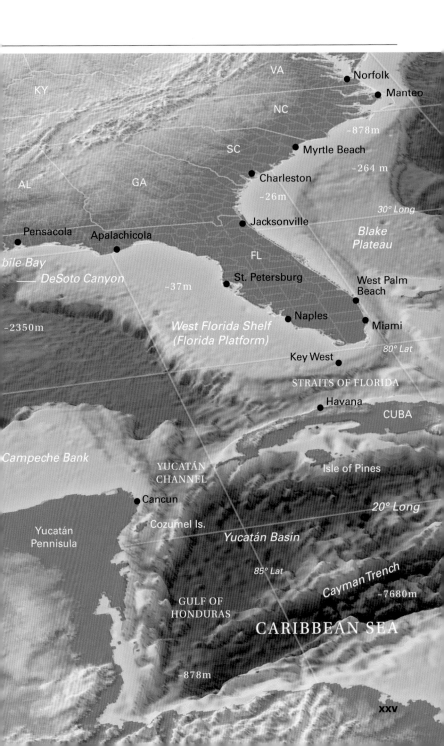

VA

KY

Norfolk

Manteo

NC

−878m

SC

Myrtle Beach

−264 m

Charleston

AL

GA

−26m

30° Long

Pensacola

Apalachicola

Jacksonville

Blake Plateau

FL

bíle Bay

____ *DeSoto Canyon*

−37m

St. Petersburg

West Palm Beach

−2350m

West Florida Shelf (Florida Platform)

Naples

Miami

Key West

80° Lat

STRAITS OF FLORIDA

Havana

CUBA

Campeche Bank

YUCATÁN CHANNEL

Isle of Pines

Cancun

20° Long

Cozumel Is.

Yucatán Basin

Yucatán Pennisula

85° Lat

Cayman Trench

−7680m

GULF OF HONDURAS

CARIBBEAN SEA

−878m

TOPOGRAPHY OF SHARKS AND RAYS

A quick overview of common topographic terminology, particularly of structures typically referred to in field descriptions

TYPICAL SHARK

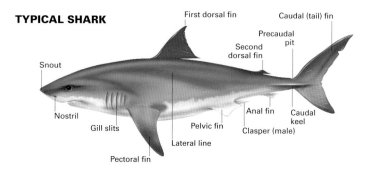

First dorsal fin · Caudal (tail) fin · Precaudal pit · Second dorsal fin · Snout · Nostril · Gill slits · Pelvic fin · Lateral line · Pectoral fin · Anal fin · Caudal keel · Clasper (male)

DOGFISH

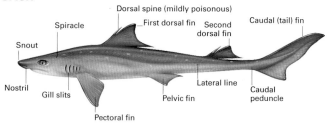

Dorsal spine (mildly poisonous) · Spiracle · First dorsal fin · Second dorsal fin · Caudal (tail) fin · Snout · Nostril · Gill slits · Pectoral fin · Pelvic fin · Lateral line · Caudal peduncle

RAY

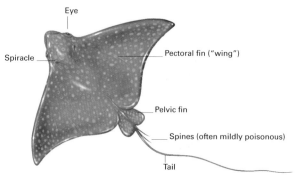

Eye · Spiracle · Pectoral fin ("wing") · Pelvic fin · Spines (often mildly poisonous) · Tail

SPINY-RAYED FISH

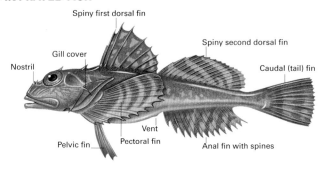

Spiny first dorsal fin

Spiny second dorsal fin

Gill cover

Nostril

Caudal (tail) fin

Vent

Pelvic fin

Pectoral fin

Anal fin with spines

TYPICAL FISH

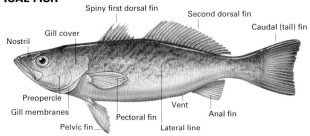

Spiny first dorsal fin

Second dorsal fin

Gill cover

Caudal (tail) fin

Nostril

Preopercle

Gill membranes

Vent

Pelvic fin

Pectoral fin

Anal fin

Lateral line

TUNA

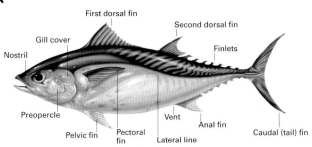

First dorsal fin

Second dorsal fin

Gill cover

Finlets

Nostril

Preopercle

Vent

Anal fin

Pelvic fin

Pectoral fin

Lateral line

Caudal (tail) fin

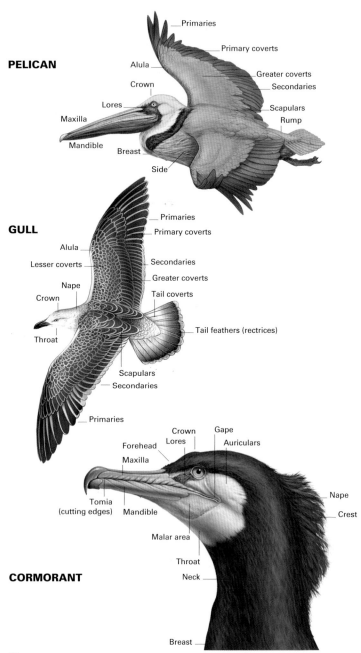

PELICAN

Primaries

Primary coverts

Alula

Crown

Greater coverts

Secondaries

Lores

Scapulars

Maxilla

Rump

Mandible

Breast

Side

GULL

Primaries

Primary coverts

Alula

Lesser coverts

Secondaries

Nape

Greater coverts

Crown

Tail coverts

Throat

Tail feathers (rectrices)

Scapulars

Secondaries

Primaries

Crown

Gape

Lores

Auriculars

Forehead

Maxilla

Nape

Crest

Tomia
(cutting edges)

Mandible

Malar area

Throat

CORMORANT

Neck

Breast

TOPOGRAPHY OF TURTLES, DOLPHINS, AND WHALES

SEA TURTLE

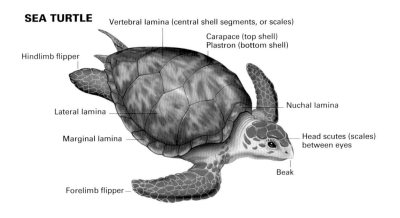

Vertebral lamina (central shell segments, or scales)

Carapace (top shell)
Plastron (bottom shell)

Hindlimb flipper

Nuchal lamina

Lateral lamina

Head scutes (scales)
between eyes

Marginal lamina

Beak

Forelimb flipper

TYPICAL DOLPHIN

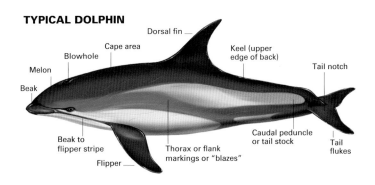

Dorsal fin

Cape area

Keel (upper
edge of back)

Blowhole

Tail notch

Melon

Beak

Beak to
flipper stripe

Caudal peduncle
or tail stock

Tail
flukes

Flipper

Thorax or flank
markings or "blazes"

HUMPBACK WHALE

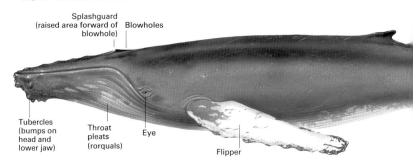

Splashguard
(raised area forward of   Blowholes
blowhole)

Tubercles
(bumps on
head and
lower jaw)

Throat
pleats
(rorquals)

Eye

Flipper

A Field Guide to the

SOUTHEAST COAST
& GULF OF MEXICO

PLATES

GULFWEED

Seaweeds, or marine algae, are associated mainly with inshore rocks, shallow bottoms, and fringes of offshore islands. The only true marine (pelagic) group of seaweeds are the Gulfweeds (*Sargassum* spp.). These free-floating mats, as the name implies, are most often found in the warm Gulf Stream waters. The algal clumps are host to a complex community of invertebrates and fish, such as the amazingly camouflaged Sargassumfish. Because of the abundance of life associated with seaweeds, birds and fish find these dense mats very attractive as feeding areas. Mats of Gulfweed can grow to vast size in the waters off Bermuda, in the becalmed region known as the Sargasso Sea.

GULFWEED
Sargassum natans and *S. fluitans*

True pelagic ocean drifters. Note small air bladders on branching tips and leaflike flattened blades. Separate the two species based on their air bladders. *S. natans* has small spikes at bladder tips; *S. fluitans* lacks this feature. Gulfweed thrives in warm waters but can drift northward in large mats. **Range:** Worldwide, throughout temperate and tropical ocean waters. **Size:** To 24 in. (61 cm).

SARGASSUMFISH
Histrio histrio

A small fish usually found well offshore, embedded in mats of Gulfweed (*Sargassum* spp.) algae. Thoroughly camouflaged by markings; almost invisible in Gulfweed. **Range:** Open ocean, but sometimes blown inshore within floating mats of algae. **Size:** To 8 in. (20 cm).

SARGASSUM TRIGGERFISH
Xanthichthys ringens

A small fish with highly variable body color that can be changed at will. Mostly tan to yellow, with lines of fine spots across flanks. Three lines below eye. Young live in Gulfweed mats; older fish live near ocean bottom in deeper water. **Range:** Cape Cod to Florida, Bahamas, Gulf of Mexico, and Caribbean. **Size:** To 10 in. (25 cm).

SLENDER SARGASSUM SHRIMP
Latreutes fucorum

Also called Gulfweed Shrimp. A *tiny* shrimp often found in great numbers within or around floating mats of Gulfweed. Normally lives well offshore, typically in warm Gulf Stream waters. Color is highly variable, depending on surrounding environment. **Range:** Throughout the Atlantic south of Cape Cod, but much more common in warmer southern waters and the Gulf Stream. **Size:** 0.75 in. (19 mm).

Sargassumfish are highly variable in color and in the shape and extent of the fin fringes and camouflage appendages that help them disappear into mats of Gulfweed.

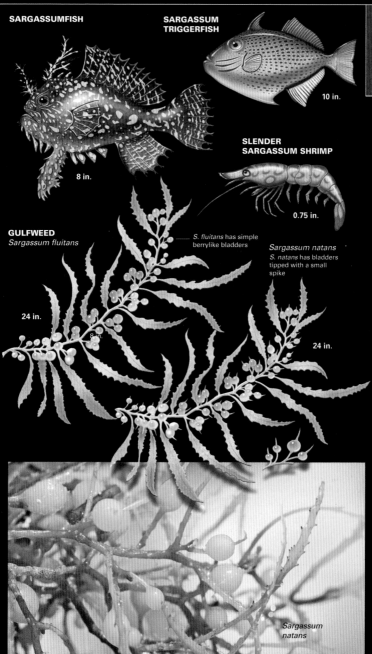

SARGASSUMFISH

SARGASSUM TRIGGERFISH

10 in.

SLENDER SARGASSUM SHRIMP

8 in.

0.75 in.

GULFWEED
Sargassum fluitans

S. fluitans has simple berrylike bladders

Sargassum natans
S. natans has bladders tipped with a small spike

24 in.

8 in.

24 in.

Sargassum natans

N. PROCTOR

3

MARINE ALGAE

GREEN HORSETAILS / PIPE-CLEANER ALGAE — *Batophora oerstedii*

Feathery, brackish-water algae that generally grow on the stilt roots of the Red Mangrove. Look for yellow-green "foxtail"-like stipes attached to roots and waving in the water. Some submerged plants have much stiffer stipes, hence the alternate name Pipe-Cleaner Algae. **Range:** South of Cape Hatteras and Gulf Coast. **Size:** To 24 in. (61 cm).

MERMAID'S WINEGLASS — *Acetabularia calyculus*

Beautiful small algae that grow in the shallow limestone waters of bays. Limestone coats their surfaces, making them stiff to the touch. Easily identified by their small, groove-capped cups atop long, thin stipes. Look for them along the bottom in shallow water, attached to rocks or, in marinas, on underwater docks or posts. **Range:** Florida, Bahamas, and Caribbean. **Size:** To 1.5 in. (3.8 cm).

PINECONE ALGAE — *Rhipocephalus phoenix*

If you enjoy snorkeling, look for this distinctive, common algae anchored to the bottom near reefs and limestone outcrops. Attractive and unmistakable, these bright green, small pinecones grow on short stipes. After storms they can be found along the wrack line. Most common in south Florida waters and farther south. **Range:** Florida, Bahamas, and Caribbean. **Size:** To 4 in. (10 cm).

FLAT-TOP BRISTLE BRUSH — *Penicillus pyriformis*

This genus has many species ranging from greenish to dull whitish in color, especially if covered with limestone deposits. In the hand it is very stiff, and its shape makes it easy to identify. Look for this algae in shallow bays among sea grasses or attached to the rocks in limestone-bottomed harbors. **Range:** Cape Hatteras to Florida, Bahamas, and Gulf Coast. **Size:** To 5 in. (13 cm).

MERMAID'S FAN — *Udotea flabellum*

A smooth, firm blade found in limestone waters, on coastal rocks, or among sea grasses. If you walk out on shelving rocks at low tide, you can find these flattened algae with a rather stiff stipe anchored to the bottom. Their flattened surface is often split along the edge from the wave action of the tides. Distinctive brownish "annual ring"–like bands mark the fan. **Range:** Cape Hatteras south to Florida, Gulf Coast, Bahamas, and Caribbean. **Size:** To 8 in. (20 cm).

COMMON DISC ALGAE — *Halimeda opuntia*

Most *Halimeda* species are distinguished by flat, nearly circular segments that attach to each other at a single basal point, giving them a jointed appearance. Segments look like flat pads of the Prickly Pear Cactus. These offshore algae of shallow to fairly deep waters can be found in the wrack line after storms. **Range:** Cape Hatteras to Florida, Bahamas, Gulf Coast, and Caribbean. **Size:** To 8 in. (20 cm).

GREEN HORSETAILS

24 in.

MERMAID'S WINEGLASS

1.5 in.

PINECONE ALGAE

4 in.

5 in.

FLAT-TOP BRISTLE BRUSH

COMMON DISC ALGAE

8 in.

MERMAID'S FAN

8 in.

5

MARINE ALGAE

HOOKED WEED   *Hypnea musciformis*

Distinct hooked, tipped branches make this algae fairly easy to identify. Color varies from yellow-brown to dull red to almost black. Common among coastal rocks exposed at low tide, as well as tangled in wrack lines where its hooked tips entwine with other algae. **Range:** Cape Hatteras south to Florida, Bahamas, Gulf Coast, and Caribbean. **Size:** To 8 in. (20 cm).

LARGE SEA GRAPES   *Botryocladia occidentalis*

Large seaweed with distinctive grapelike clusters of bladders (to 0.2 in., 5 mm, long) on branches. Dull rose to yellow-brown in color. Found in limestone-based waters up to 65 ft. (20 m) deep. Underwater, it resembles a small fruited tree and is interesting to observe while snorkeling. Often found in the wrack line after heavy surf. **Range:** Cape Hatteras south to Florida, Bahamas, Gulf Coast, and Caribbean. **Size:** To 10 in. (25 cm).

RED WEED   *Agardhiella tenera*

Extremely common along the entire coastline. Large clumps can be found washed up on the beach or tangled in the algae "rolls" left in the wrack line after high tide. Feels somewhat gelatinous to the touch. **Range:** Cape Hatteras south to Florida, Bahamas, Gulf Coast, and Caribbean. **Size:** To 12 in. (30 cm).

SEA LETTUCE   *Ulva lactuca*

Very common and widespread in coastal waters. Bright green with leaflike texture. Sheets may reach 2 ft. (61 cm) or more in size. Can be abundant in nitrogen-rich waters. One of the most common algae to wash up on the shoreline. **Range:** Labrador to Florida, Bahamas, Gulf Coast, and Caribbean. **Size:** To 24 in. (61 cm).

GREEN FLEECE   *Codium isthmocladum*

Once you feel this velvetlike branched algae, the common name becomes obvious. The many species of *Codium* all have rounded branches, a velvet texture, and a well-developed holdfast at the base of the main stipe. *Codium* is the bane of the oystermen, as it anchors to shells of oysters and causes currents to pull the oysters off the oyster beds. They can grow large (to 3 ft., 91 cm, across), but more often are smaller and left behind on the beach after storms or high tides. **Range:** Cape Hatteras south to Florida, Bahamas, Gulf Coast, and Caribbean. **Size:** To 36 in. (91 cm).

PETTICOAT ALGAE   *Padina vickersiae*

Flat, fan-shaped blades with incurved margins and distinct "growth ring"–like lines on the fans identify this algae. Found in bays, sea grass beds, and limestone-based waters, where it anchors to the bottom. Not often seen in wrack lines. An interesting algae to see while snorkeling. **Range:** Cape Hatteras south to Florida, Bahamas, Gulf Coast, and Caribbean. **Size:** To 4 in. (10 cm).

HOOKED WEED

8 in.

LARGE SEA GRAPES

10 in.

RED WEED

12 in.

SEA LETTUCE

24 in.

GREEN FLEECE 36 in.

PETTICOAT ALGAE 4 in.

MARINE GRASSES

Sea grass beds provide critical shelter and food resources for a wide variety of coastal marine animals. Unfortunately, sea grasses are particularly susceptible to pollution, ocean warming, and the diseases that accompany these environmental changes.

EELGRASS
Zostera marina

Classic grasslike green leaves. Usually grows in beds well below the low-tide line, though sometimes also in the intertidal zone. The most common sea grass through the Carolinas and Georgia. Much less common south of Georgia. **Range:** Nova Scotia to Georgia. **Size:** Blade length varies widely, with longer leaves in deeper waters.

TURTLE GRASS
Thalassia testudinum

Wide green blades. The most common marine grass in the Florida Keys and along the Gulf Coast. Turtle Grass beds appear as dark patches offshore, and dead Turtle Grass commonly lines Florida beaches. **Range:** South Florida, Gulf Coast, and Caribbean. **Size:** Blades highly variable in length, and about 1 in. (2.5 cm) in width.

MANATEE GRASS
Syringodium filiforme

Long, slender round leaves. Often grows interspersed in grass flats with Turtle Grass and marine algae such as Mermaid's Fan (*Udotea* sp.). **Range:** Gulf of Mexico, south Florida, and Bahamas. **Size:** Blade length varies, 6–15 in. (15–38 cm), longest in deeper waters.

Lush Turtle Grass beds just off the beach at Bahia Honda State Park, Bahia Honda Key, Florida. Note the accumulation of dry sea grass along the tide lines on the beach.

P. LYNCH

EELGRASS

NOAA

Kemp's Ridley
Sea Turtle

TURTLE GRASS

MIXED TURTLE AND MANATEE GRASSES

NOAA

West Indian
Manatee

MANATEE GRASS
AND MERMAID'S
FAN ALGAE

Mangal or mangrove communities form dense tangled forests and marshes along tropical shores the world over. Mangroves are unique in their ability to root in poorly oxygenated tidal mud flats, where their roots are constantly exposed to salt water.

The dense underwater roots of mangroves provide shelter for the young of many marine species, and the rich variety of animal and plant life that mangroves support both above and below the water's surface are crucial in providing food for the offshore coral reefs and grass flats of Florida Bay and the Florida Keys.

Pictured here are Red Mangroves, the most salt-tolerant of the three common mangrove species. Just behind the shoreline screen of Red Mangroves, Black Mangroves are found on slightly higher ground, while White Mangroves grow on the driest ground.

John Pennekamp Coral Reef State Park, Key Largo, Florida

Red Mangroves
Rhizophora mangle

P. LYNCH

MANGROVES

BLACK MANGROVE
Avicennia germinans

Favors shallower waters than Red Mangroves, but often mixes with Red and White Mangroves. Vertical root extensions called pneumatophores are diagnostic, as is the dark, reticulated bark on larger trees. Leaves are often covered with salt crystals. **Range:** Florida and Texas coasts, most Caribbean islands. **Size:** Small bushes to trees 30 ft. (9.1 m) tall.

RED MANGROVE
Rhizophora mangle

Classic mangrove of sea islands in the tropics. The arching prop roots are instantly recognizable, as are the long, pointed seedlings that drop arrowlike into the mud to form new trees. **Range:** South Florida, Florida Keys, south Texas, and most Caribbean islands. **Size:** Small bushes to trees 30 ft. (9.1 m) tall.

WHITE MANGROVE
Laguncularia racemosa

Most upland and least cold-tolerant mangrove, often mixing with Buttonwoods in mangrove salt marshes and sea islands. Rounded oval leaves, with small glands at base of leaf stem and along underside leaf margins. **Range:** Central Florida through Florida Gulf Coast and Texas; Caribbean islands. **Size:** Mostly low shrubs, but can rise to 30 ft. (9.1 m).

BUTTONWOOD
Conocarpus erectus

A mangrove relative that favors drier ground in mangrove swamps and marshes, often mixed with White Mangroves. Elongated, pointed leaves, with distinct leaf and stalk glands. Named for its rounded, white flower clusters. **Range:** Central and southern Florida and Caribbean islands. **Size:** Mostly low shrubs, but mature trees can rise to 30 ft. (9.1 m) or more.

A young Red Mangrove showing the characteristic arching prop roots.

Black Mangroves are surrounded by vertical roots extensions called pneumatophores.

P. LYNCH

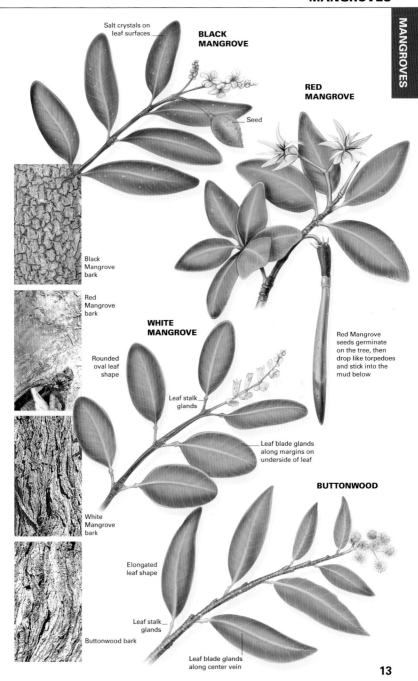

BLACK MANGROVE

Salt crystals on leaf surfaces

Seed

RED MANGROVE

Black Mangrove bark

Red Mangrove bark

Red Mangrove seeds germinate on the tree, then drop like torpedoes and stick into the mud below

WHITE MANGROVE

Rounded oval leaf shape

Leaf stalk glands

Leaf blade glands along margins on underside of leaf

White Mangrove bark

BUTTONWOOD

Elongated leaf shape

Leaf stalk glands

Buttonwood bark

Leaf blade glands along center vein

13

MANGROVE SALT MARSHES

Merritt Island National Wildlife Refuge is near the northern limit of the Black and White Mangroves. Here the two species mix with marsh grasses and coastal trees to form a unique mangrove marsh habitat that once stretched unbroken south along Florida's east coast to the Florida Keys.

Vertical root branches called pneumatophores surround the base of Black Mangroves

White Mangrove
Laguncularia racemose

White Mangroves line the drier edges of a dike road through the salt marsh

MANGROVE SALT MARSHES

On higher ground, Slash Pines, Cabbage Palms, and Buttonwoods mix into the mangroves

Black Mangroves typically line the deeper creeks in the marsh, while White Mangroves grow on slightly higher ground behind

Black Mangrove
Avicennia germinans

**Mangrove marsh
Merritt Island
National Wildlife Refuge
Titusville, Florida**

MANGROVE SALT MARSH PLANTS

BLACK MANGROVE
Avicennia germinans

Northernmost mangrove. Look for pneumatophores, or "breathing roots," at base to confirm species from a distance. **Range:** East coast of Florida almost to Georgia border in small numbers and west coast of Florida north to Texas Panhandle coast. **Size:** In Florida averages 4–10 ft. (1.2–3 m) as a shrub, but can reach tree size in tropics.

WHITE MANGROVE
Laguncularia racemosa

Very similar to Black Mangrove at a distance (see pg. 13), but look for the more rounded leaves with clusters of light green seedpods. **Range:** Cape Canaveral and Tampa south and south Texas coast; at scattered locations in small numbers elswhere along Gulf Coast. **Size:** In Florida, averages 4–15 ft. (1.2–4.6 m) as a shrub, but can reach tree size in tropics.

BROOMSEDGE
Andropogon virginicus

Tall grass with bushy crown. Very common along roadsides and on disturbed ground. Tolerates salt on barrier islands and in coastal marshes. **Range:** Throughout Southeast and Gulf states. **Size:** To 3 ft. (0.9 m).

CABBAGE PALMETTO
Sabal palmetto

Common Florida palm with rounded crown. Trunk usually crisscrossed with stumps of old leaf bases. **Range:** Throughout Florida; hugs coastline north to Cape Hatteras. **Size:** To 80 ft. (24.4 m).

SLASH PINE
Pinus elliottii

Florida's most common coastal pine. Mixes with Loblolly and Longeaf Pines along coastlines north and west of Florida. **Range:** Throughout Florida, north along coast to Georgia, and west to Louisiana. **Size:** To 100 ft. (30 m), though rarely that tall along coast.

Typical Florida coastal wetlands. Along the waterways Black and White Mangroves mix with Smooth and Saltmeadow Cordgrasses. On higher ground Cabbage Palmettos, taller White Mangroves, Buttonwood, mixed Slash, Loblolly, and Longleaf Pines predominate. In tropical south Florida, the Gumbo-Limbo Tree (*Bursera simaruba*) enters the mix on higher ground.

P. LYNCH

BLACK MANGROVE PNEUMATOPHORES

BLACK MANGROVE SEEDPODS

WHITE MANGROVE SEEDPODS

BROOMSEDGE

CABBAGE PALMETTO

SLASH PINE

P. LYNCH

Loblolly Pine
Pinus taeda

Longleaf Pine
Pinus palustris

Smooth Cordgrass
Spartina alterniflora

Black Needlerush
Juncus roemerianus

The southeastern coastal salt marshes form mostly on the landward sides of barrier islands and on the mainland along the shores of more sheltered bays, sounds, and tidal rivers. The ecology and species composition of salt marshes is largely determined by the extent of tidal flooding by salt water.

In tidal salt marshes the marsh is flooded twice daily, and the flora is dominated by the most salt-tolerant grasses, such as Smooth and Saltmeadow Cordgrass. In tidal brackish marshes that receive more freshwater from rivers, the species mix includes more Black Needlerush, Spike Grass, and Narrow-Leaved Cattails. Behind the marshes, on higher ground, grow the shrubs and trees of the maritime Loblolly Pine and Live Oak forests typical of the higher ground of southeastern barrier islands and coastal plains.

Nags Head Woods Ecological Preserve, Kill Devil Hills, North Carolina

P. LYNCH

The maritime forest:
Dogwood, *Cornus florida*
Laurel Oak, *Quercus laurifolia*
Live Oak, *Quercus virginiana*
Eastern Red-Cedar, *Juniperus virginiana*
Yaupon Holly, *Ilex vomitoria*

Smooth Cordgrass
Spartina alterniflora

SALT MARSH AND MARITIME FOREST PLANTS

SMOOTH CORDGRASS   *Spartina alterniflora*

Also known as Saltwater Cordgrass. A common and dominant species of salt
marsh grass along the East and Gulf Coasts. Tall; usually seen at seaward edges of
salt marshes, where water is most saline. **Range:** Newfoundland south throughout
East and Gulf Coasts. **Size:** To 4–5 ft. (1.2–1.5 m).

SALTMEADOW CORDGRASS   *Spartina patens*

Also called Salt Hay. A dominant component of salt marshes throughout East
and Gulf Coasts. Short; forms on slightly higher ground than the closely related *S.
alterniflora*. Can form low marsh meadows that stretch for miles along the coast.
Range: Newfoundland south throughout East and Gulf Coasts. **Size:** To 1–2 ft.
(30–61 cm).

BLACK NEEDLERUSH   *Juncus roemerianus*

In coastal marshes Black Needlerush marks the transition from a purely saltwater
marsh to marshes with a mix of salt and freshwater. Typically lines the waterway
edges of brackish estuary marshes, and may form extensive tall "meadows"
behind barrier islands close to rivers or brackish sounds. **Range:** Southern New
Jersey to Florida and Gulf Coast. **Size:** To 2–5 ft. (0.3–1.5 m).

LOBLOLLY PINE   *Pinus taeda*

Dominant pine along Carolina and Georgia coasts. Often forms mixed stands
with very similar Longleaf Pine in barrier island maritime forests dominated by
Live Oak. **Range:** Maryland to northern Florida and Florida Panhandle west to
Louisiana. **Size:** To 100 ft. (30 m).

LIVE OAK   *Quercus virginiana*

The majestic Live Oak virtually defines southeastern maritime and coastal forests.
In the maritime forests of barrier islands it is often the most numerous large tree
and, when mature, is certainly the most noticeable. The classic mature Live Oak
has a short central trunk and massive branches that make the crown far wider
than the tree is tall. **Range:** Southeastern Virginia to Florida and entire Gulf Coast.
Size: To 65 ft. (20 m).

The huge spreading crown of a mature Live Oak is unmistakable at a distance. These trees
are in the Outer Banks at Currituck, North Carolina.

P. LYNCH

SMOOTH CORDGRASS

SALTMEADOW CORDGRASS

BLACK NEEDLERUSH

LOBLOLLY PINE

LIVE OAK, LEAVES

LIVE OAK

P. LYNCH

This grouping of Saw Palmetto, Sea Grape, and Sea Oats is native to Florida and Caribbean beaches, but these hardy plants are now also widely planted to control erosion on Florida beaches.

Playalinda Beach, Cape Canaveral National Seashore, Florida

P. LYNCH

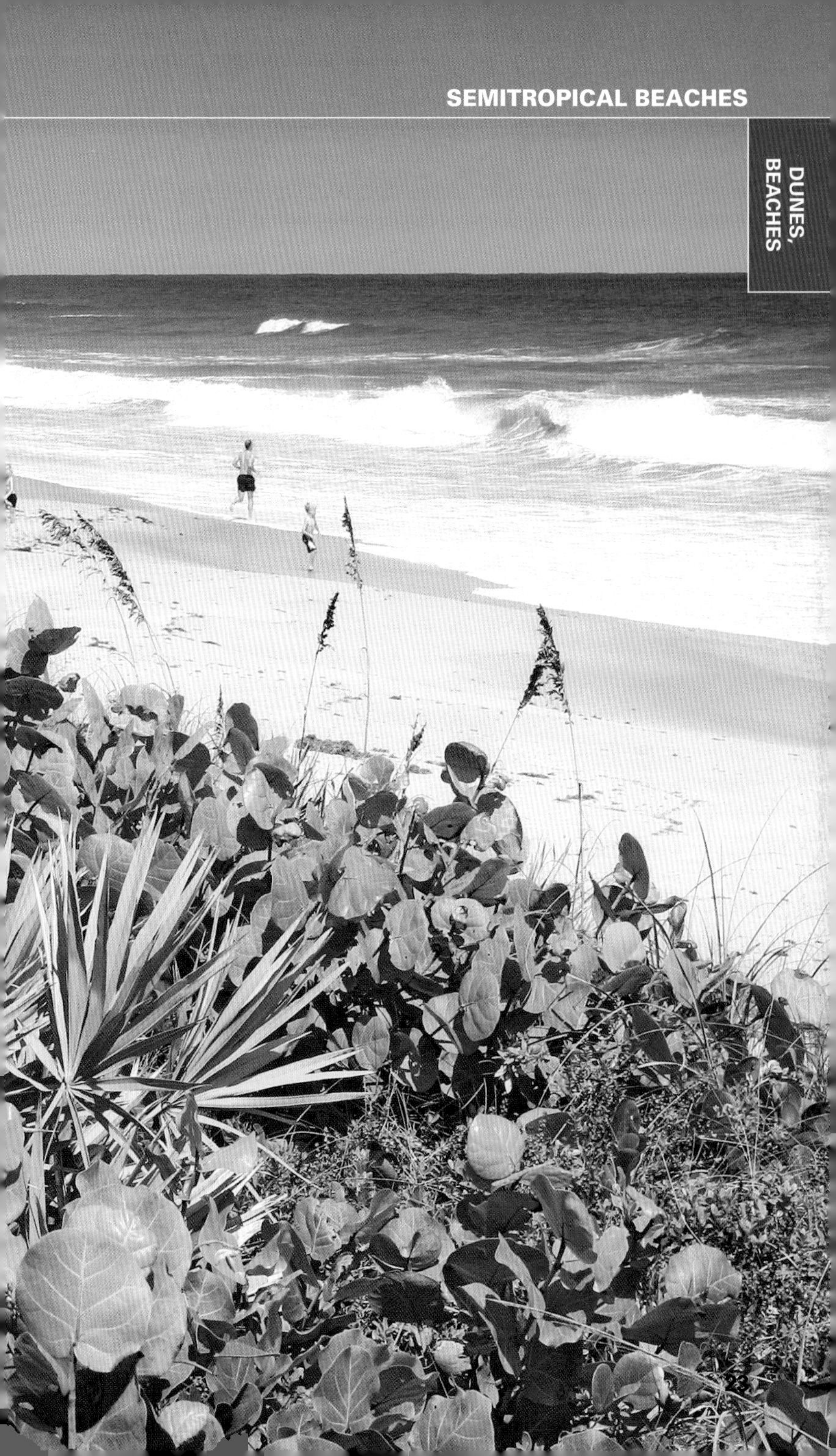

SEMITROPICAL BEACH PLANTS

SAW PALMETTO
Serenoa repens

Very common, distinctive ground cover and low shrub on coastal islands and beaches. Can reach small tree size in maritime forests. Widely planted to control beach erosion. **Range:** South Carolina to Florida and Gulf Coast. **Size:** Typically to 3–6 ft. (0.9–1.8 m).

SEA GRAPE
Coccoloba uvifera

Classic tropical beach plant of the Caribbean and south Florida, now also widely planted on Florida beaches to control erosion. Unmistakable, with large, leathery oval leaves. **Range:** Central Florida beaches south to Florida Keys, Bahamas, and much of the Caribbean. **Size:** Mostly a low shrub, but in Caribbean reaches tree size (50 ft., 15.2 m).

SEA OATS
Uniola paniculata

This large, tough grass is often the land plant closest to the surf on beaches. Sea Oats are critical to erosion control all along the Atlantic and Gulf Coasts, as their roots quickly stabilize loose sand. **Range:** Southeastern Virginia to Florida, Bahamas, and entire Gulf Coast. **Size:** To 4 ft. (1.2 m).

PRICKLY PEAR CACTUS
Opuntia humifusa

Most beaches are dry as deserts, so it's no surprise to find the common Prickly Pear Cactus in the sand, but usually in back dune areas well away from direct salt spray or salt water soaking the sand. **Range:** Massachusetts to Florida and entire Gulf Coast. **Size:** To 36 in. (91 cm) in height.

CABBAGE PALMETTO
Sabal palmetto

Common Florida palm. Rounded crown over trunk that is usually crisscrossed with stumps of old leaf bases. Found near beaches, but away from direct salt spray, typically behind first or second set of coastal dunes. **Range:** Throughout Florida, hugging coastline north to Cape Hatteras. **Size:** To 80 ft. (24.4 m), but coastal trees are much smaller.

SLASH PINE
Pinus elliottii

Florida's most common coastal pine. Mixes with Loblolly and Longleaf Pines along coastlines north and west of Florida. **Range:** Florida, north along coast to Georgia, and west to Louisiana. **Size:** To 100 ft. (30 m), although rarely that tall along the coast.

DUNES, BEACHES

SAW PALMETTO

SEA GRAPE

SEA OATS

PRICKLY PEAR CACTUS

CABBAGE PALMETTO

SLASH PINE

P. LYNCH

Sea Oats
Uniola paniculata

Railroad Vine
Ipomoea pes-caprae

Sandbur
Cenchrus tribuloides

Beach Croton
Croton punctatus

Russian Thistle
Salsola kali

P. LYNCH

Seabeach Orache
Atriplex cristata

Sea Oats
Uniola paniculata

The hardy plants that occupy the narrow strip of land between the ocean and inland are remarkably adapted to the constant onslaught of drying winds and wind-blown salt.

Plants that can survive here have evolved a unique range of adaptations. A thick cutin layer on the leaves reduces water loss and lessens abrasion from wind-blown sand. Deep roots anchor the plant and bring water from well below the surface. Branching runners spread across the sand, helping both to spread the plant and to stabilize the sand below it.

The plants shown here are only a fraction of the dune species you might see, but all will show a similar range of adaptations to this demanding coastal environment.

BEACH AND DUNE PLANTS

AMERICAN BEACH GRASS
Ammophila breviligulata

The most common beach grass on the East Coast. Planted to control erosion south of Cape Hatteras, and critical for stabilizing beaches and dunes. **Range:** Newfoundland to Florida. **Size:** To 14 in. (36 cm).

SEA OATS
Uniola paniculata

Replaces American Beach Grass as the common native beach grass south of Cape Hatteras, but due to plantings both species are commonly mixed south to Florida. **Range:** Virginia to Florida and Gulf Coast. **Size:** 4 ft. (1.2 m).

GLASSWORTS
Salicornia sp. and *Sarcocornia* sp.

Common in salt marshes and on beaches. Annual *Salicornia* glassworts range from Massachusetts south to Cape Hatteras. The similar perennial glasswort *Sarcocornia* ranges from Cape Hatteras south and throughout Gulf Coast. **Size:** Typically a low creeper on sandy soils, under 12 in. (30 cm) tall.

WAX MYRTLE
Myrica cerifera

Tough, salt-resistant shrub common on back dunes shielded from direct salt spray. Leathery leaves are fragrant when torn. **Range:** New Jersey south to Florida and Gulf Coast through Louisiana. **Size:** Typically 3–5 ft. (0.9–1.5 m).

POISON IVY
Toxicodendron radicans

Poisonous; produces a painful rash if touched. "Leaves of three, let it be." Very common on dunes and at salt marsh edges. **Range:** Nova Scotia to Florida and entire Gulf Coast. A vine that climbs whatever support is available for height.

VIRGINIA CREEPER
Parthenocissus quinquefolia

Very common climbing and spreading vine on back dunes, sandy areas, and marsh edges, mixed with other vegetation. **Range:** Maine to Florida and entire Gulf Coast. A climbing vine that also extends along the ground.

Sea Oats (*Uniola paniculata*) at Cape Hatteras, North Carolina

P. LYNCH

DUNES, BEACHES

AMERICAN BEACH GRASS

SEA OATS

GLASSWORT

WAX MYRTLE

POISON IVY

VIRGINIA CREEPER

JELLYFISH

Siphonophores and jellyfish are two groups that are often confused and lumped as one group. However, they are distinctly different animals. Siphonophores are hydrozoans, which have an attached, treelike (hydra) stage and a free-floating (medusa) stage. Siphonophores exhibit an extreme form of this combination: they are complex colonies of medusae and polyps. The Portuguese Man-Of-War is an excellent example. Siphonophores also have a velum (sail) because they are planktonic (weak swimmers) and drift with the wind. Jellyfish (Scyphozoa) are large marine medusae whose polyp stage is either lacking or reduced to an extremely small form. True jellyfish are free-swimming and lack a velum.

BY-THE-WIND SAILOR
Velella velella

Small platform structure with brownish, flattened central sail (velum). Below disk, mass of tentacle-like structures up to 5 in. (12.5 cm) long, with feeding and reproductive flaps. A true marine form. Rapidly disintegrates once it drifts inshore and water salinity drops. **Range:** Tropical waters worldwide. **Float size:** To 4 in. (10 cm).

PORTUGUESE MAN-OF-WAR
Physalia physalia

Well known for its painful stings. Powder blue, balloonlike transparent float is visible at the water's surface. Elongate tentacle-like processes and curtainlike folds hang below inflated float and may extend more than 50 ft. (15.2 m). *Beware of tentacles, which can cause severe stings and rashes, even long after animal's death.* **Range:** Throughout tropical and warm temperate waters worldwide. **Float size:** To 1 ft. (30 cm).

SEA NETTLE
Chrysaora quinquecirrha

Large jellyfish with smooth to slightly pebbly bell surface. Tentacles emerge from distinct clefts on bell edge. Two distinct forms. Marine form is larger (to 7 in., 18 cm), with more tentacles (up to 40) and with pink lines on bell. Estuarine and bay form is smaller (to 4 in., 10 cm), with 24 tentacles. *Sting is very painful, causing severe burns and rashes.* **Range:** Cape Cod south to West Indies. Very common in Chesapeake Bay and other tidal areas. Water salinity determines range limits.

SEA WASP
Tamoya haplonema

Rigid bell with four distinct tentacles, each with flattened, paddlelike base and elongated filament. Strong, fast swimmer that tends to stay near the sea bottom. *Sting is extremely painful; use caution.* **Range:** Mainly tropical; on occasion drifts as far north as Long Island. **Float size:** To 4 in. (10 cm).

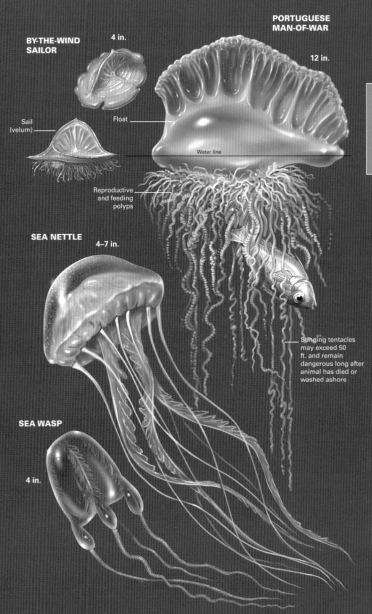

BY-THE-WIND SAILOR

4 in.

PORTUGUESE MAN-OF-WAR

12 in.

Sail (velum)

Float

Water line

Reproductive and feeding polyps

SEA NETTLE

4–7 in.

Stinging tentacles may exceed 50 ft. and remain dangerous long after animal has died or washed ashore

SEA WASP

4 in.

31

MOON JELLY
Aurelia aurita

Familiar jellyfish of bays, sounds, and inland waters. Bell edge rimmed by short fringe of tentacles. Prey captured in mucilaginous cap and "cleared off" for ingestion by elongate mouth arms that drape below. Easily identified by the distinct shamrock appearance of gonads seen through the transparent cap. *Can cause an itchy rash or mild sting if handled.* **Range:** Extreme northern Greenland to Caribbean. **Size:** To 10 in. (25 cm) diameter.

LION'S MANE JELLYFISH
Cyanea capillata

The world's largest jellyfish. A classic jellyfish, ranging in color from brownish to pink. Extremely long tentacles and massive cascading group of mouth lobes. *Contact with tentacles can cause severe stings and burns.* **Size:** Typically to 12 in. (30 cm) diameter, but may reach 8 ft. (2.4 m). Most specimens south of Cape Hatteras are small (to 5 in., 12.5 cm), but individuals increase in size in North Atlantic waters. **Range:** North Atlantic to the Carolinas.

MUSHROOM JELLYFISH
Rhopilema verrilli

Cream colored with a thick spherical bell. Brownish yellow marks on mouth lobes. the Mushroom Jellyfish does not have true tentacles, but has long, finger-like feeding appendages. Mainly marine, but can be blown into sounds and bays under certain wind conditions. *May cause a mild sting or rash if handled.* **Range:** Tropical seas. Less common than other jellies in our waters, but can be locally abundant off Carolinas and south to Florida and the Gulf. A rare stray north of Cape Hatteras. **Size:** To 20 in. (51 cm) diameter, but more typically 10–12 in. (25–30 cm).

MOON JELLY
Typically 8–10 in.
diameter

Tentacles

Gonads

Oral arms

LION'S MANE JELLYFISH

Typically to 12 in.
diameter, but can
reach huge sizes,
up to 8 ft.

MUSHROOM JELLYFISH
Typically 10–12 in.
diameter

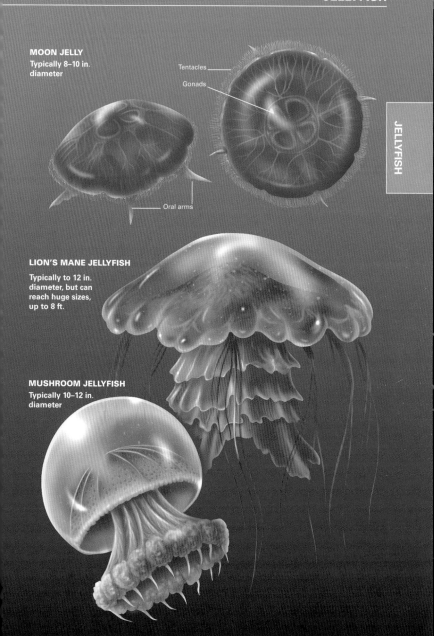

CANNONBALL AND UPSIDE-DOWN JELLIES

CANNONBALL JELLYFISH
Stomolophus meleagris

Named for its near-spherical bell. Does not sting humans, but toxins in its body may cause skin irritations if it is handled. Abundant in coastal waters in summer and fall. **Range:** Wanders north to Long Island, but most common south of Cape Hatteras, throughout Caribbean and West Indies, and south to Brazil. Often seen in huge swarms off the Gulf Coast. **Size:** Typically 4–8 in. (10–20 cm) diameter.

UPSIDE-DOWN JELLYFISH
Cassiopeia frondosa

Very common in mangrove swamps, shallow bays, and lagoons. Much less common in sandy beach areas. Typically rests on the bottom and swims "upside-down," with tentacles on top. *Causes an itchy rash or painful welt if handled.* **Range:** Common along south Florida coasts and throughout Caribbean in suitable coastal habitats. **Size:** Typically 4–5 in. (10–13 cm) diameter.

MANGROVE UPSIDE-DOWN JELLYFISH
Cassiopeia xamachana

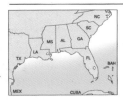

Very common in mangrove swamps, marshes, Turtle Grass flats, and other areas where coastal plant communities meet the sea. Much less common in sandy beach areas. Typically rests on the bottom and swims "upside-down," with tentacles on top. *Causes an itchy rash if handled.* **Range:** Common along south Florida coasts and throughout Caribbean in coastal habitats. **Size:** To 7–8 in. (18–20 cm) in diameter.

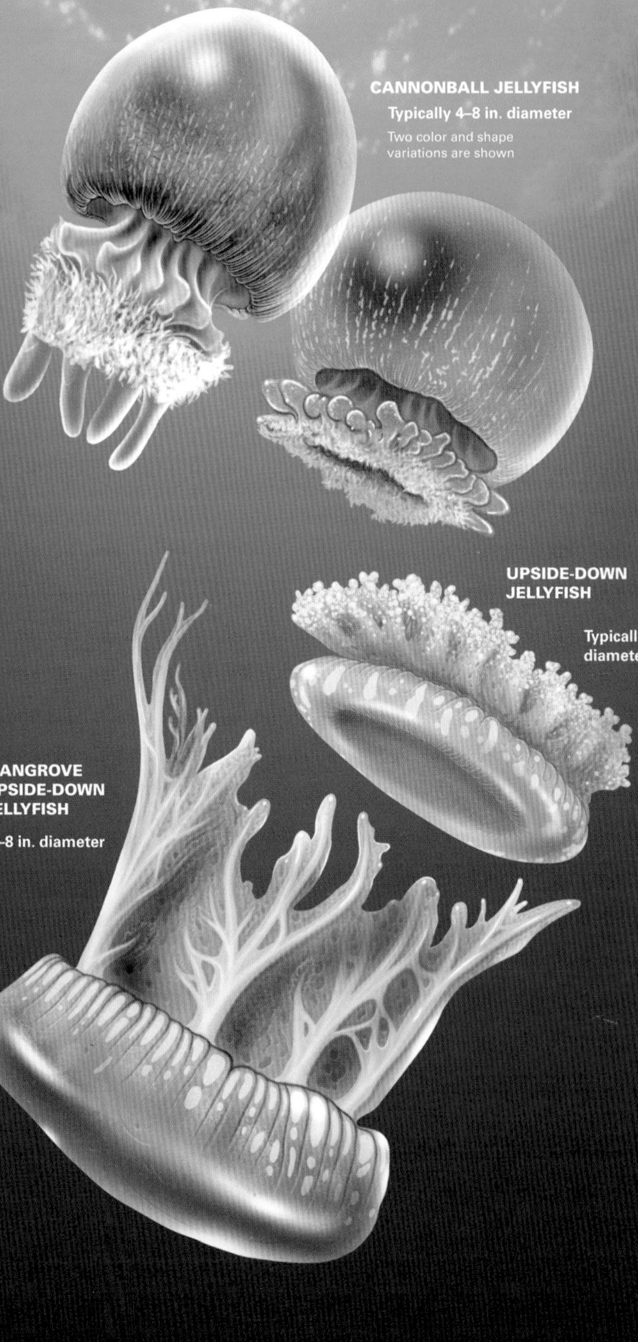

CANNONBALL JELLYFISH

Typically 4–8 in. diameter

Two color and shape
variations are shown

**UPSIDE-DOWN
JELLYFISH**

**Typically 4–5 in.
diameter**

**MANGROVE
UPSIDE-DOWN
JELLYFISH**

7–8 in. diameter

COMB JELLIES (CTENOPHORES)

Often mistaken for jellyfish. Comb jellies, however, do not sting, have two tentacles or lobes below their saclike body, and have up to eight comblike ciliary plates. Gelatinous and fragile, they turn into nondescript blobs out of water. All are plankters (drifters with little self-locomotion) and can amass in large swarms. They are predators of small fish and fish eggs. *Note:* When viewing comb jellies, look for pinkish wormlike forms in the gut region. These are the parasitic young of the Burrowing Anemone (*Edwardsia leidyi*).

SEA GOOSEBERRY   *Pleurobrachia pileus*

Walnut-shaped body. Commonly drifts into inshore waters and washes up on beaches, where contractile tentacles cannot be seen. Incredible numbers can be washed into bays along the mid-Atlantic Coast. Unusual south of South Carolina. Colorless and hard to see in water. **Size:** To 1.2 in. (2.8 cm).

NORTHERN COMB JELLY   *Bolinopsis infundibulum*

Lobes shorter than oval body. The most common comb jelly in northern Atlantic coastal waters; occasionally found south to Cape Hatteras. Transparent and hard to see in water. Like all comb jellies, can occur in amazingly large numbers. **Size:** To 6 in. (15 cm).

LEIDY'S COMB JELLY (SEA WALNUT)   *Mnemiopsis leidyi*

Body more flattened than Northern Comb Jelly. Lobes longer than body sphere. When disturbed, bright green bioluminescent flashes occur all along combs. The most common comb jelly along the southeastern Atlantic Coast and Gulf of Mexico. **Size:** To 4 in. (10 cm).

BEROE'S COMB JELLY   *Beroe ovata*

A transparent, somewhat flattened sac. Pinkish to rust colored with comb rows on bell and no tentacles. Found from Virginia south and in Gulf of Mexico. **Size:** To 4.5 in. (11 cm).

VENUS GIRDLE   *Cestum veneris*

Unmistakable Gulf Stream species. Flattened like a belt with central constriction. Two rows of combs on upper lobe margins. Greenish to white in color. Rare drifter out of Gulf Stream to southeastern coastal waters. **Size:** To 5 ft. (1.5 m).

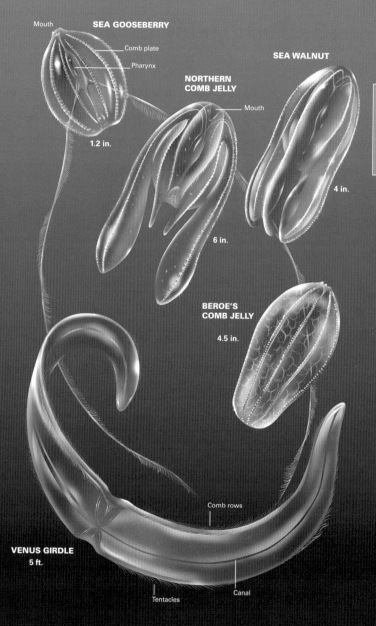

Mouth

SEA GOOSEBERRY

Comb plate

Pharynx

SEA WALNUT

**NORTHERN
COMB JELLY**

Mouth

1.2 in.

4 in.

6 in.

**BEROE'S
COMB JELLY**

4.5 in.

Comb rows

VENUS GIRDLE
5 ft.

Tentacles

Canal

Common Sea Fan
Gorgonia ventalina

Scrawled Filefish
Aluterus scriptus

P. LYNCH

Stoplight Parrotfish
Sparisoma viride

Rough Sea Plume
Muriceopsis flavida

Corky Sea Fingers
Briareum asbestinum

Molasses Reef
John Pennekamp Coral Reef State Park
Key Largo, Florida

BRANCHING AND BRAIN CORALS

ELKHORN CORAL
Acropora palmata

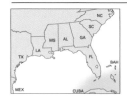

Large and branching, with thick, flattened branches reminiscent of elk antlers. Once one of the most prominent reef-building corals, Elkhorn has suffered a 95-97% decline since 1980 due to coral bleaching and hurricane damage. **Range:** Florida Keys and most of Caribbean. **Size:** Colonies 3-12 ft. (0.9-3.7 m). **Red List – Critically endangered**

STAGHORN CORAL
Acropora cervicornis

Antlerlike hard coral with slim, cylindrical branches ranging from a few inches to over 6.5 ft. (2 m) in length. Once a dominant coral in the Florida Keys, Staghorn has suffered a 95-97% decline since 1980 due to coral bleaching and hurricane damage. **Range:** Florida Keys and most of Caribbean. **Size:** Colonies 1-8 ft. (0.3-2.4 m). **Red List – Critically endangered**

GROOVED BRAIN CORAL
Diploria labyrinthiformis

The most common brain coral in most reefs, typically growing in large, globular "brainlike" colonies. **Range:** Common throughout Caribbean at depths of 20-80 ft. (6-24.4 m). **Size:** Colonies 1-4 ft. (0.3-1.2 m) in diameter.

BOULDER BRAIN CORAL
Colpophyllia natans

One of the most common brain corals, sometimes called Giant Brain Coral. **Range:** Common throughout Caribbean at depths of 20-80 ft. (6-24.4 m). **Size:** Colonies 1-8 ft. (0.3-2.4 m).

MAZE CORAL
Meandrina meandrites

Colonies may form globular "brain" formations or various flat plates or irregular encrustations. **Range:** Common throughout Caribbean at depths of 2-200 ft. (0.6-61 m) **Size:** Colonies 1-3 ft. (0.3-0.9 m).

SYMMETRICAL BRAIN CORAL
Diploria strigosa

Colonies form convoluted "brain" patterns in a variety of shapes ranging from classic globular "brains" to flat plates and irregular encrustations. **Range:** Common throughout Caribbean at depths of 3-120 ft. (0.9-37 m). **Size:** Colonies 1-6 ft. (0.3-1.8 m).

BRAIN CORAL PATTERNS

BOULDER BRAIN CORAL

GROOVED BRAIN CORAL

MAZE CORAL

SYMMETRICAL BRAIN CORAL

ELKHORN CORAL

STAGHORN CORAL

GROOVED BRAIN CORAL

41

PILLAR, STAR, AND FIRE CORALS

PILLAR CORAL
Dendrogyra cylindrus

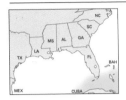

Mature colonies form tall, irregular "pillars" with many fingerlike vertical projections. **Range:** Common throughout Caribbean south of Florida at depths of 3–120 ft. (0.9–37 m). Less common but still frequently seen in reefs off Florida Keys, where it is susceptible to coral bleaching and has suffered a reduction in population in recent decades. **Size:** Colonies 3.5–10 ft. (1–3 m) in height. **Red List – Vulnerable**

MOUNTAINOUS STAR CORAL
Montastraea faveolata

Typically forms large mounds with short, conelike vertical projections covering colony surface. Formerly a dominant reef-forming hard coral, Mountainous Star Coral has declined 50% in recent decades due to coral bleaching. Still common in Florida Keys reefs. **Range:** Throughout Caribbean at depths of 5–120 ft. (1.5–37 m). **Size:** Colonies up to 10 ft. (3 m) diameter. **Red List – Endangered**

BLADED FIRE CORAL
Millepora complanata

Fire corals, though not true corals, are a common component of reef communities in the Caribbean, south Florida, and some Gulf areas. This species typically forms flat blades or fans, often mustard yellow or yellow-brown, that make it easy for divers to spot. *Produces a painful sting and fiery itch if touched or rubbed against.* **Range:** Common throughout south Florida and Caribbean waters at depths of 1–40 ft. (0.3–12 m). **Size:** Colonies to 20 in. (51 cm) diameter.

SMOOTH STAR CORAL
Solenastrea bournoni

Typical star coral, named for the stony, star-shaped cups that the live coral polyps sit within on the coral's surface. Colonies of Smooth Star Coral form in a wide variety of shapes, including globular mounds and large, encrusted plates. **Range:** One of the most cold-tolerant of the stony corals, ranging from small colonies as far north as the Carolinas to larger colonies commonly seen in Florida Keys and Gulf reefs. Unusual south of Florida and Gulf of Mexico. **Size:** Colonies typically 1–3 ft. (0.3–0.9 m) diameter.

PILLAR, STAR, FIRE CORALS

PILLAR CORAL
3.5–10 ft. high

MOUNTAINOUS STAR CORAL
To 10 ft. diameter

BLADED FIRE CORAL
To 20 in. diameter

STAR CORAL
To 36 in. diameter

SOFT CORALS (GORGONIANS)

Often mistaken for marine plants, these soft corals are closely related to reef-forming corals, but they do not form the limestone skeletons that support the hard corals, such as Staghorn, Elkhorn, Brain, and Star Corals.

COMMON SEA FAN
Gorgonia ventalina

Sea fans form colonies of flat, branched fan blades positioned for maximum exposure to water currents flowing over the reef. This is the most common sea fan in south Florida and Florida Keys reefs. Color varies but always includes much purple. **Range:** Bermuda, throughout Caribbean and south Florida, but largely absent from Gulf of Mexico. **Size:** 2–3 ft. (0.6–0.9 m) at depths of 4–100 ft. (1.2–30 m), but may reach 6 ft. (1.8 m).

BENT SEA ROD
Plexaura flexuosa

Sea rods are colonies of polyps that may grow in either of two dominant patterns: as bushy, free-form branches (shown here) or as branches that align on a flat plane. Color varies from tan or yellowish brown to reddish purple. **Range:** Common throughout Bahamas, south Florida, and Caribbean. **Size:** 6–16 in. (15–41 cm) in height at depths of 3–150 ft. (0.9–46 m).

CORKY SEA FINGERS
Briareum asbestinum

Colonies of erect cylinders or rods. Appearance varies depending on whether the coral polyps are extended (thick, hairy appearance) or contracted (more fingerlike, smooth rods). See illustration at right for variations. **Range:** Abundant in south Florida and Bahama reefs and throughout Caribbean. **Size:** 3–24 in. (8–61 cm) at depths of 3–100 ft. (0.9–30 m).

BRANCHING TUBE SPONGE
Pseudoceratina crassa

Common tubular sponge with wide variety of growth habits and forms, ranging from tall, cylindrical tubes to shorter, barrel-like forms. Surface may be relatively smooth or very bumpy, and colors range from yellow and red to dark green and purple. **Range:** Common in Florida Keys and Bahama reefs; increasingly abundant in southern Caribbean and Antilles. **Size:** 6–20 in. (15–51 cm) at depths of 20–85 ft. (6.1–26 m).

**COMMON
SEA FAN**
2–3 ft. high

BENT SEA ROD
6–16 in. high

**CORKY SEA
FINGERS**
Polyps contracted

**CORKY SEA
FINGERS**
Polyps extended

BRANCHING TUBE SPONGE

45

PELAGIC INVERTEBRATES (PLANKTON)

SHELLED SEA BUTTERFLY
Limacina spp.

Tiny snail-like organism that sometimes gathers in vast numbers near the ocean surface. Usually seen well offshore at night in deep water at the surface. Shell transparent. Swims by paddling with winglike parapodia (modified snail feet). **Size:** 0.25 in. (6 mm).

NAKED SEA BUTTERFLY
Clione limacina

Similar in appearance and habits to Shelled Sea Butterfly but larger and lacking hard shell. Paired tentacles on head and three pairs of tentacle-like projections. Swims using winglike parapodia (modified snail feet). **Size:** 1 in. (25 mm).

ARROW WORM
Sagitta spp.

Very small predatory worm that swims like a tiny fish at the ocean surface, preying on small plankton. Hooked claws and tiny eyespots on head. Two paired fins along length of body, flattened tail. Generally found well offshore. **Size:** 0.75 in. (19 mm).

PLANKTON WORM
Tomopteris helgolandica

Glassy, transparent planktonic worm with centipede-like appearance. Active swimmer, seen at night at the surface of offshore waters. Curls into a ball and sinks quickly when touched or frightened. **Size:** To 3.5 in. (89 mm).

OIKOPLEURA
Oikopleura spp.

Tiny invertebrates that create gelatinous bubblelike "glass houses" about 0.75 in. (19 mm) in diameter. Look closely for tiny tadpolelike animal within this glass house. Often very common in coastal waters. **Size** (of animal): 0.5 in. (13 mm).

SALP
Thalia and *Salpa* spp.

Transparent jellylike plankton, similar in size and shape to a gel medicine capsule, with an open central core through body. Moves by jetting water through body. Can be present in huge numbers, generally well offshore but sometimes in coastal waters. **Size:** Varies with species, from less than 0.25 in. (6 mm) to 8.5 in. (21 cm).

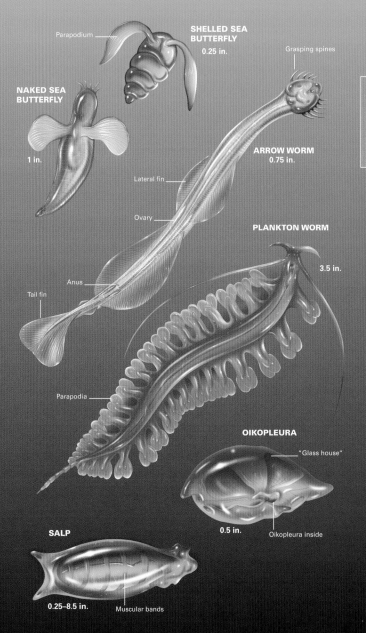

Parapodium

SHELLED SEA BUTTERFLY
0.25 in.

Grasping spines

NAKED SEA BUTTERFLY

1 in.

ARROW WORM
0.75 in.

Lateral fin

Ovary

PLANKTON WORM

3.5 in.

Anus

Tail fin

Parapodia

OIKOPLEURA

"Glass house"

0.5 in.

Oikopleura inside

SALP

0.25–8.5 in.

Muscular bands

SHELLS OF BEACHES AND SANDY COASTS

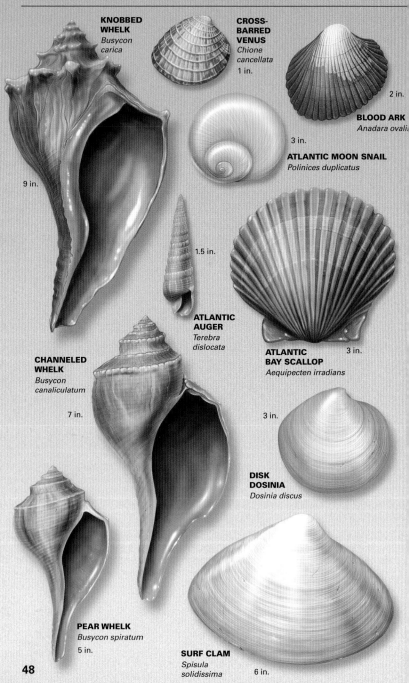

KNOBBED WHELK
Busycon carica

9 in.

CROSS-BARRED VENUS
Chione cancellata
1 in.

BLOOD ARK
Anadara ovali
2 in.

ATLANTIC MOON SNAIL
Polinices duplicatus
3 in.

1.5 in.

ATLANTIC AUGER
Terebra dislocata

ATLANTIC BAY SCALLOP
Aequipecten irradians
3 in.

CHANNELED WHELK
Busycon canaliculatum
7 in.

DISK DOSINIA
Dosinia discus
3 in.

PEAR WHELK
Busycon spiratum
5 in.

SURF CLAM
Spisula solidissima
6 in.

SHELLS OF BEACHES AND SANDY COASTS

ANGELWING
Cyrtopleura costata
5 in.

LETTERED OLIVE
Oliva sayana
2.5 in.

LIGHTNING WHELK
Busycon contrarium
5 in.

LACE MUREX
Chicoreus dilectus
2 in.

GIANT ATLANTIC COCKLE
Dinocardium robustum
4 in.

COMMON JANTHINA
Janthina janthina
1 in.

EASTERN OYSTER
Crassostrea virginica
3 in.

CALICO SCALLOP
Aequipecten gibbus
2 in.

SOUTHERN QUAHOG
Mercenaria campechiensis
5 in.

SQUID, ARGONAUT (CEPHALOPODS)

An active and intelligent group of animals. Note the complex eyes, which rival those of mammals. The head and body of a squid is termed the mantle. Fin shape is very important in the identification of the more common squid.

CARIBBEAN REEF SQUID
Sepioteuthis sepioidea

The common squid of coral reefs and inshore areas in the tropics. Swims by rippling fins that run the length of the body on either side. Color is highly variable. When alarmed, or when displaying to another squid, can change color rapidly, from pale sand through iridescent shades to almost black. **Range:** South Florida, Bahamas, most of Gulf of Mexico, Caribbean. **Size:** Mantle 6–12 in. (15–30 cm).

ATLANTIC BRIEF SQUID
Lolliguncula brevis

Distinctly rounded fins are less than half the length of the mantle. Feeds on a wide array of small fish. **Range:** Inshore from Delaware Bay south. **Size:** Mantle to 5 in. (13 cm).

LONGFIN INSHORE SQUID
Loligo pealeii

Fins very long to half the length of mantle and angled. No eye notch. Eaten by a wide variety of fish. The most common open-water squid south of Cape Hatteras. **Range:** Bay of Fundy south to Caribbean. **Size:** Mantle to 17 in. (43 cm).

GREATER ARGONAUT (PAPER NAUTILUS)
Argonauta argo

Elegant shelled drifter of deep seas. Best known through its beautiful "paper shell," which washes up on beaches, actually an egg case secreted by the female's modified arms. The male is shell-less. **Range:** Warm tropical waters north to Cape Cod. **Size:** Mantle to 12 in. (30 cm).

BLUE GLAUCUS
Glaucus atlanticus

A sea snail, not a cephalopod. Semitransparent body is silvery to deep blue. Invertebrate drifter in the ocean's upper layer, often among Gulf Weed and other ocean debris. A predator of small jellyfish and other plankton. Often found in debris and floating vegetation. **Range:** Northern Atlantic. **Size:** Length to 1.5 in. (3.8 cm).

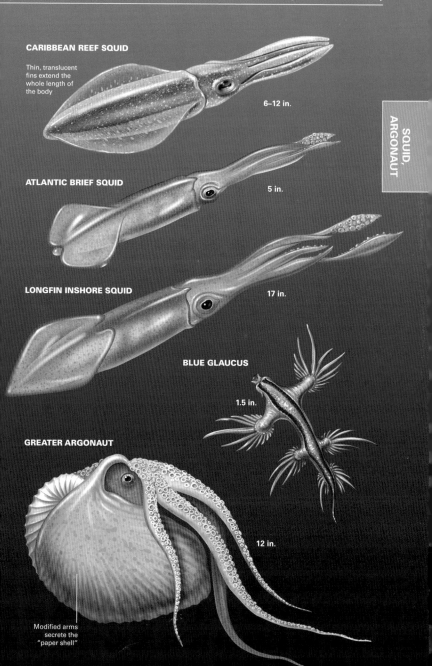

CARIBBEAN REEF SQUID

Thin, translucent
fins extend the
whole length of
the body

6–12 in.

ATLANTIC BRIEF SQUID

5 in.

LONGFIN INSHORE SQUID

17 in.

BLUE GLAUCUS

1.5 in.

GREATER ARGONAUT

12 in.

Modified arms
secrete the
"paper shell"

SHRIMP

The three major *Penaeus* edible species are not easily identified to species level, as they are all very similar in general body form, and all three species are highly variable in color. All *Peneaus* shrimp have antennae longer than their bodies, and have tiny claws on the first three legs on each side.

WHITE SHRIMP
Penaeus setiferus

Whitish blue or gray. Live shrimp have pink sides. Also called Northern White Shrimp or Gray Shrimp. Common edible shrimp species along the southeastern Atlantic Coast in waters less than 100 ft. (30 m) deep. **Range:** Long Island south to Florida Keys, Gulf of Mexico. **Size:** To 7 in. (18 cm).

BROWN SHRIMP
Penaeus aztecus

Brown to olive-green; sometimes entirely dark red or green. Common edible shrimp species along the southeastern Atlantic Coast in waters less than 100 ft. (30 m) deep. **Range:** Cape Cod south to Florida Keys, Gulf of Mexico. **Size:** To 7 in. (18 cm).

PINK SHRIMP
Penaeus duorarum

Normally pinkish or grayish white, but varies to red-brown or gray-brown. Also called Northern Pink Shrimp. Common edible shrimp species along the southeastern Atlantic Coast in waters less than 100 ft. (30 m) deep. **Range:** Cape Cod south to Florida Keys, Gulf of Mexico. **Size:** To 7 in. (18 cm).

SLENDER SARGASSUM SHRIMP
Latreutes fucorum

Also called Gulfweed Shrimp. A *tiny* shrimp often found in great numbers within or around floating mats of Gulf Weed. Normally lives well offshore, typically in warm Gulf Stream waters. Sargassum Shrimp are often blown ashore within large clumps of *Sargassum*. Exact color is highly variable, depending on surrounding environment. **Range:** Throughout Atlantic south of Cape Cod, but much more common in warmer southern waters and Gulf Stream. **Size:** 0.75 in. (19 mm).

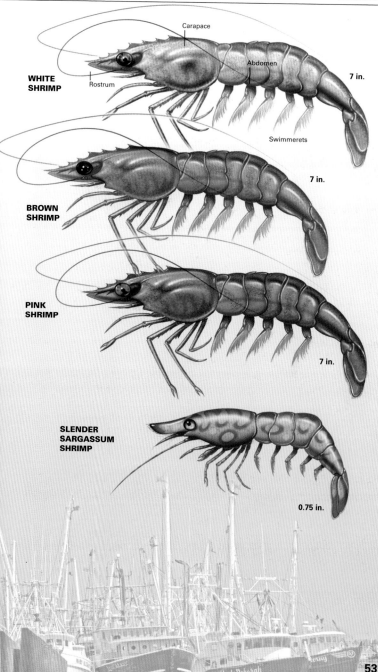

WHITE
SHRIMP

Carapace

Rostrum

Abdomen

Swimmerets

7 in.

BROWN
SHRIMP

7 in.

PINK
SHRIMP

7 in.

SLENDER
SARGASSUM
SHRIMP

0.75 in.

SHRIMP

LOBSTERS

NORTHERN LOBSTER
Homarus americanus

Familiar to anyone who has visited a seafood restaurant, with two massive claws on the first legs and long, slender antennae almost as long as the body. Color varies from dark brown to red-brown; very rarely blue. **Range:** Mainly Labrador to New Jersey, but small populations are found south to North Carolina. **Size:** To 36 in. (91 cm), but averaging 12–16 in. (30–41 cm).

CARIBBEAN SPINY LOBSTER
Panulirus argus

The familiar edible lobster south of Cape Hatteras. Very spiny carapace and head. No large claws; massive second antennae are longer than body. Usually brown or olive green overall, with prominent lighter stripes on the legs and spotted patterns on the sides and tail. Nocturnal, but often seen peeking out of sheltered holes in reefs by divers. **Range:** Cape Hatteras south to Brazil, Bahamas, Caribbean, and Gulf of Mexico. **Size:** To 24 in. (61 cm), but normally 8–10 in. (20–25 cm).

SPANISH LOBSTER
Scyllarides aequinoctialis

Also called Spanish Slipper Lobster. Red-brown in color with lighter mottled areas across carapace and tail. The antennae are large, flat plates projecting from the head. No large claws. **Range:** South Florida, Bahamas, and throughout Caribbean at depths of 25–120 ft. (7.6–37 m). **Size:** To 6–12 in. (15–30 cm).

The Caribbean Spiny Lobster as typically seen by divers, backed into a small cave in the reef. Spiny lobsters are primarily nocturnal and are usually secretive and wary during the day.

P. LYNCH

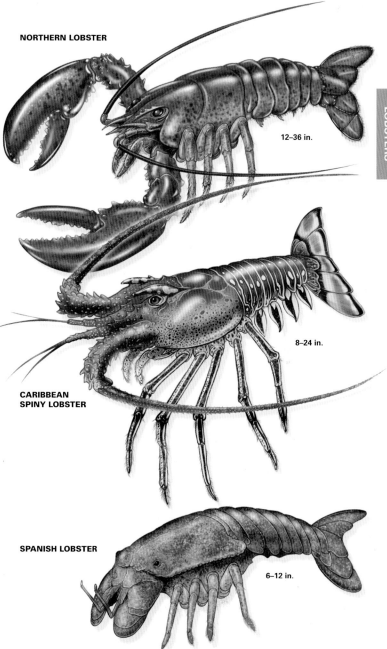

NORTHERN LOBSTER

12–36 in.

**CARIBBEAN
SPINY LOBSTER**

8–24 in.

SPANISH LOBSTER

6–12 in.

CRABS

BLUE CRAB

Callinectes sapidus

Common in many shoreline habitats from near-shore waters to marshes and brackish tidal rivers and on all types of bottoms from sand to rocks or mud. Spindle-shaped body is dark brown or green, with bright blue legs and claws. A strong, fast swimmer. Very aggressive when cornered: use gloves when handling. Very important commercial seafood, particularly in mid-Atlantic states. **Range:** Massachusetts Bay to Florida, Caribbean, Gulf Coast, and south to Uruguay. **Size:** Carapace to 9 in. (23 cm) wide.

STONE CRAB

Menippe mercenaria

Stout, reddish gray to brownish gray, mottled with fine spots. Prominent claws, usually marked with violet-red areas; dark claw tips. The largest of the mud crab family, usually found well below the low tide line hiding in burrows, rocks, or heavy vegetation on the bottom. Such a popular South Florida seafood that many people think this crab's first name is "Joe's." **Range:** North Carolina south to Florida Keys, Bahamas, and West Indies, and Gulf of Mexico south to Yucatán. **Size:** Carapace to 4.5 in. (11 cm) wide.

LADY CRAB

Ovalipes ocellatus

Perhaps our most beautiful crab. Warm yellow-brown carapace marked with a fine pattern of tiny dark red circles. Legs usually marked with bright red-orange and yellow areas and dark spots. Like the Blue Crab, the Lady Crab is a strong swimmer well known for its aggressive disposition and tendency to draw blood if handled casually. **Range:** Cape Cod to Georgia. **Size:** Carapace to 3.25 in. (8.3 cm) wide.

BLUE CRAB

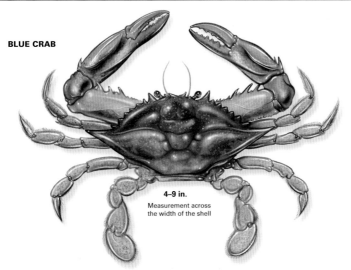

4–9 in.
Measurement across
the width of the shell

STONE CRAB

Juvenile

Adult

0.5–1 in.

4.5 in.

LADY CRAB

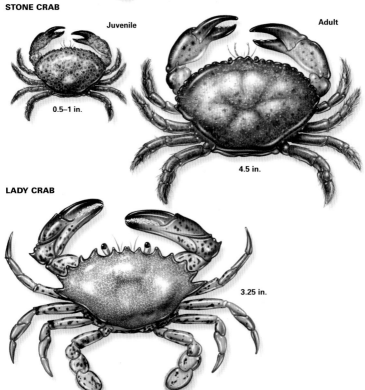

3.25 in.

57

CRABS

SAND FIDDLER CRAB
Uca pugilator

The most common of several fiddler crab species, which frequent mud flats and muddy-sandy beaches in marshes and bays sheltered from ocean waves. Burrows in small holes, where it hides from the sun and predators during low tides. **Range:** Cape Cod to Florida and Gulf Coast to Texas. **Size:** Carapace to 1 in. (2.5 cm) wide.

GHOST CRAB
Ocypode quadrata

The most common and visible crab on sandy ocean beaches and dunes, where they dig numerous burrows to escape predators and the hot sun of the day. Very active as beach scavengers at night. **Range:** Rhode Island to Florida, Gulf Coast, and Bahamas south to Brazil. **Size:** Carapace to 2 in. (5 cm) wide.

JONAH CRAB
Cancer borealis

Stout, heavy-bodied, dark brown or reddish brown and covered with fine spotted patterns. Most common on rocky shores and bottoms, where it ranges from the low tide line into deep offshore waters. **Range:** Nova Scotia to Florida. **Size:** Carapace to 6 in. (15 cm) wide.

ATLANTIC ROCK CRAB
Cancer irroratus

Common, red-brown or yellow-brown, with fine darker spotting across the carapace. Similar to the Jonah Crab, but with a proportionately smaller carapace and more prominent legs and claws. **Range:** Labrador to Georgia. **Size:** Carapace to 5 in. (13 cm) wide.

Beaches are covered with Ghost Crab holes and diggings in the early morning. The holes shelter the crabs from predators and the heat of the day. For the best views of Ghost Crabs, take a flashlight out onto the beach at night.

P. LYNCH

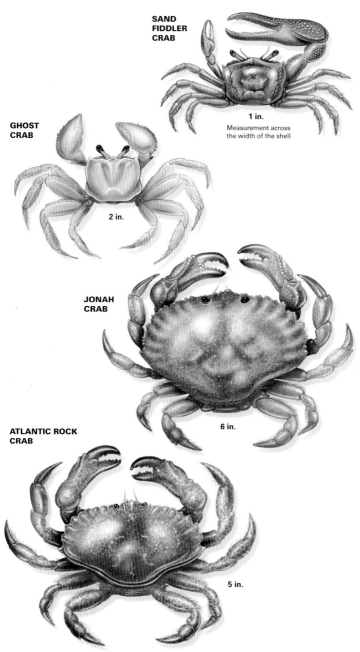

SAND FIDDLER CRAB

1 in.
Measurement across the width of the shell

GHOST CRAB

2 in.

JONAH CRAB

6 in.

ATLANTIC ROCK CRAB

5 in.

CRABS

CRABS

STRIPED HERMIT CRAB
Clibanarius vittatus

Claws are *equal* in size; all other shallow-water hermit crabs have unequal claws. Medium-sized, brown, with bright yellow or greenish stripes on walking legs. **Range:** Outer Banks to Florida and Gulf Coast. **Size:** 1.5 in. (3.8 cm) long.

FLAT-CLAWED HERMIT CRAB
Pagurus pollicaris

Medium-sized, reddish or pink, with claws broadly flattened and unequal in size. In shallow waters along beaches and salt marsh shores. **Range:** Cape Cod to Florida and Gulf Coast. **Size:** To 1 in. (2.5 cm) in width.

ASIAN SHORE CRAB (*non-native species*)
Hemigrapsus sanguineus

Invasive species native to China and Japan, spreading rapidly along the Northeast and mid-Atlantic coastline and crowding out native crab species as it goes. Found in shallow intertidal areas. **Range:** Now common in New Jersey and Delaware, moving south to Carolinas in smaller populations. **Size:** Carapace to 1.5 in. (3.8 cm) wide.

COMMON SPIDER CRAB
Libinia emarginata

Large, long-legged, with round body, often covered with algae and debris. Found on all types of bottoms from very shallow waters to depths of 100 ft. (30 m) or more. Visually impressive, but sluggish and harmless. **Range:** Nova Scotia to Florida and Gulf Coast to Texas. **Size:** Carapace to 4 in. (10 cm) wide; leg span to 1 ft. (30 cm).

HORSESHOE CRAB
Limulus polyphemus

Unmistakable. Not a true crab, and more closely related to spiders. Very visible along the shore in midspring, when it mates and lays eggs in intertidal zones of beaches and mud flats. **Range:** Gulf of Maine to Florida and Gulf of Mexico. **Size:** 1–2 ft. (30–61 cm).

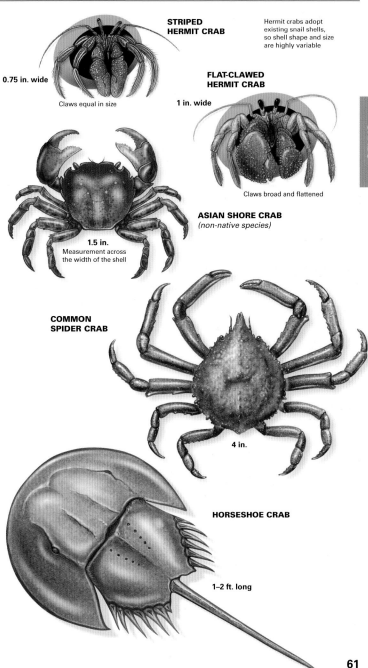

STRIPED HERMIT CRAB

Hermit crabs adopt existing snail shells, so shell shape and size are highly variable

0.75 in. wide

Claws equal in size

FLAT-CLAWED HERMIT CRAB

1 in. wide

Claws broad and flattened

CRABS

ASIAN SHORE CRAB
(non-native species)

1.5 in.
Measurement across the width of the shell

COMMON SPIDER CRAB

4 in.

HORSESHOE CRAB

1–2 ft. long

WHALE SHARK
Rhincodon typus

Description: The largest living fish. Huge gray, bluish, or greenish shark with mottled pattern of light spots on back and flanks. Three pronounced ridges run along both sides of back, merging into prominent caudal keels on either side of tail. Head very wide, with terminal mouth. **Red List – Vulnerable**

Habits: A slow-swimming, surface filter feeder. Though uncommon throughout our area, it is easily observed if found. Usually solitary, but may aggregate into groups in tropical waters. Like the Basking Shark, it feeds just below the water's surface, but its dorsal and tail fins do not often break the surface. Generally docile; often ridden by divers who grab onto its dorsal fins or back ridges. Sometimes reported to ram boats; more likely, such boats rammed the sharks.

Range: New York south through Caribbean; warmer temperate and tropical waters worldwide. Sporadic reports of sightings north of Cape Cod.

Size: Averages 25–30 ft. (7.6–9.1 m). Largest measured specimen 40 ft. (12.2 m) and 26,594 lbs. (12,063 kg), but there are numerous reliable reports of individuals reaching 60 ft. (18.3 m).

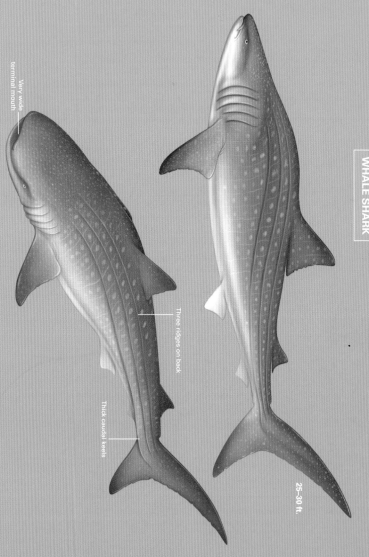

Very wide terminal mouth

Three ridges on back

Thick caudal keels

25-30 ft.

Description: Huge brown or slate-gray shark with gigantic mouth and very long gill slits that extend from its back to near midline of throat. Massive lunate tail with large caudal keels. Tiny teeth. The second largest fish after the Whale Shark. Extremely long gill slits nearly circle the head. A filter feeder like baleen whales. Special baleenlike extensions of gill arches called rakes filter planktonic animals from water. **Red List – Vulnerable**

Habits: Docile, sluggish swimmer, not known to harm swimmers or boats. Sometimes aggregates in groups and may swim head-to-toe in lines of six to eight individuals. Though reasonably common, its natural history is not well known. In winter it disappears from surface waters, moving into deep waters offshore.

Range: Newfoundland to North Carolina, possibly farther south as well; temperate waters worldwide. Prefers cold, upwelling waters where plankton is richest, migrating south along the East Coast in winter, then north again in summer.

Size: Huge; a 30-ft. (9.1 m) specimen can weigh 8,600 lbs. (3,900 kg). Basking Sharks may reach 32 ft. (9.7 m) in length, but most individuals in our area average 22–29 ft. (6.7–8.8 m) or smaller.

BASKING SHARK

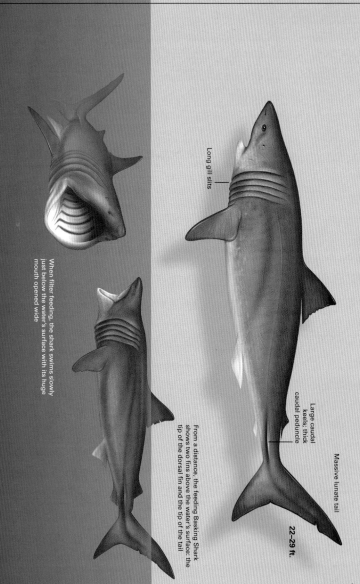

When filter feeding, the shark swims slowly just below the water's surface with its huge mouth opened wide

Long gill slits

From a distance, the feeding Basking Shark shows two fins above the water's surface: the tip of the dorsal fin and the tip of the tail

Large caudal keels; thick caudal peduncle

Massive lunate tail

22–29 ft.

MACKEREL SHARKS, BLUE SHARK

WHITE SHARK
Carcharodon carcharias

Conical snout, unmarked dorsal fin, and single caudal peduncle. Back color varies from slate gray to blue or dark brown. May show black spot in axil of pectoral fin. Adults of huge girth, heavier than other sharks of similar length. Known to attack humans. **Range:** All temperate waters, but rare. **Size:** To 25 ft. (7.6 m), but most under 16 ft. (4.9 m). Red **List – Vulnerable**

PORBEAGLE
Lamna nasus

Bluish above with sharp contrast to white underparts. Short, rounded rostrum. Note double-keeled peduncle. Heavily fished for its liver. **Range:** South to Cape Hatteras in cold waters. **Size:** To 10 ft. (3 m). **Red List – Vulnerable**

SHORTFIN MAKO
Isurus oxyrinchus

Beautiful steel blue, contrasting to white underparts. Very slender with elongate snout and high upper caudal fin. Elongate peduncle keel. Pectoral fin short. Mainly pelagic. Often leaps from water in pursuit of prey. **Range:** Cape Cod to Argentina. **Size:** To 12 ft. (3.7 m). **Red List – Vulnerable**

LONGFIN MAKO
Isurus paucus

Separated from look-alike Shortfin Mako by dark underparts except for white belly area. Pectoral fin longer and more swept back. A deep-water hunter until evening, when it rises to cool surface waters. **Range:** North Carolina to Cuba. **Size:** To 12 ft. (3.7 m).

BLUE SHARK
Prionace glauca

Spectacular shimmering blue with pure white underparts. Long, pointed snout. Very long pectoral fins. Slender, graceful appearance. Known to attack humans. **Range:** Nova Scotia and Grand Banks to Argentina. **Size:** To 12.5 ft. (3.8 m). **Red List – Near threatened**

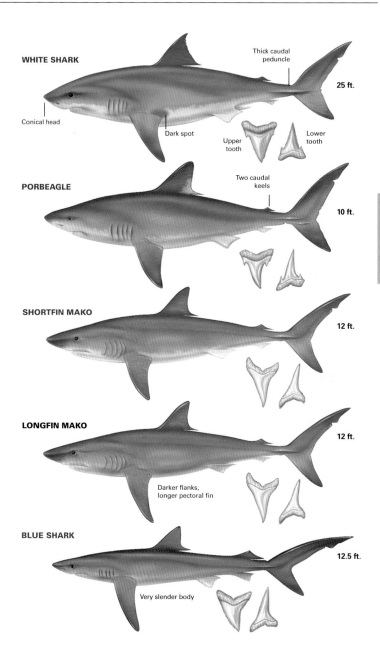

WHITE SHARK

Thick caudal peduncle

25 ft.

Conical head

Dark spot

Upper tooth

Lower tooth

PORBEAGLE

Two caudal keels

10 ft.

SHORTFIN MAKO

12 ft.

LONGFIN MAKO

12 ft.

Darker flanks, longer pectoral fin

BLUE SHARK

12.5 ft.

Very slender body

SHARKS

67

LARGE PELAGIC SHARKS

These are true pelagic species that favor deeper waters. The Tiger Shark will also come into very shallow areas. Both the Tiger Shark and the Oceanic Whitetip Shark are considered very dangerous to humans.

OCEANIC WHITETIP SHARK
Carcharhinus longimanus

Warm brown with creamy underparts. White fin tips make this shark easy to identify. A deepwater shark in temperate waters, coming in to reefs and inshore only in tropical waters. Very dangerous. **Range:** Cape Cod to Uruguay. **Size:** To 12 ft. (3.7 m). **Red List – Near threatened**

TIGER SHARK
Galeocerdo cuvier

Brownish to blue-gray with distinctive dark blotches and barring. Blunt snout. Usually pelagic and solitary. Often comes into bays at night to feed. Very dangerous. **Range:** Cape Cod to the tropics, but uncommon north of Florida. **Size:** May reach huge size, to 24 ft. (7.3 m). **Red List – Near threatened**

SILKY ("SICKLE") SHARK
Carcharhinus falciformis

Slender, bluish gray with contrasting underparts. Note fairly small dorsal. Pectoral fin very large and streamlined. Named for its remarkably smooth skin. Primarily a deepwater species, sometimes quite numerous. **Range:** Cape Cod to Brazil. **Size:** To 12 ft. (3.7 m).

BLACKTIP SHARK
Carcharhinus limbatus

Grayish blue above, pale below. Dorsal, anal, and lower caudal fins tipped with black. Inside of pectoral tips dark. White flank patch. Pelagic but often follows schools of mackerel inshore. **Range:** Cape Cod to Brazil. **Size:** To 8 ft. (2.4 m). **Red List – Near threatened**

SPINNER SHARK
Carcharhinus brevipinna

Very similar in every aspect to Blacktip Shark. But dorsal fin more to rear, snout much more elongate, and flank patch merely a white line, not a patch. Erupts from the water with spectacular spinning leaps reminiscent of Sailfish breaking the ocean surface. **Range:** North Carolina to Brazil. **Size:** To 10 ft. (3 m). **Red List – Near threatened**

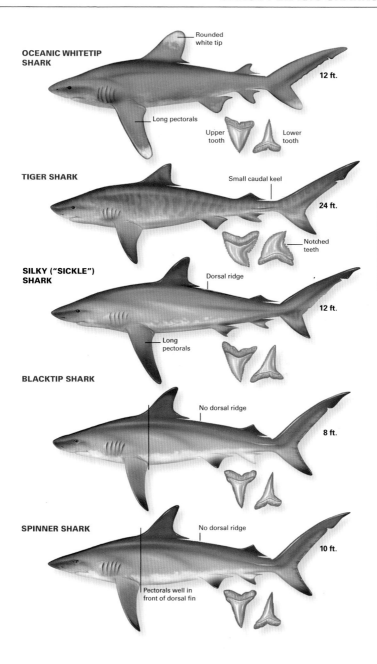

OCEANIC WHITETIP SHARK

Rounded white tip

12 ft.

Long pectorals

Upper tooth

Lower tooth

TIGER SHARK

Small caudal keel

24 ft.

Notched teeth

SILKY ("SICKLE") SHARK

Dorsal ridge

12 ft.

Long pectorals

BLACKTIP SHARK

No dorsal ridge

8 ft.

SPINNER SHARK

No dorsal ridge

10 ft.

Pectorals well in front of dorsal fin

SHARKS

PILOTFISH, REMORAS

Pilotfish are free-swimming fish that prefer to live in association with large sharks or other predatory fish. They are also commonly found under floating mats of Gulfweed. Pilotfish do not have sucker disks and never attach themselves to their hosts. Remoras and sharksuckers directly attach themselves to large sharks, Manta Rays, and large bony fish, feeding on scraps of their hosts' dinners and taking smaller fish not eaten by the host. Sharksuckers sometimes attach themselves to human swimmers—the encounters are startling but harmless.

PILOTFISH   *Naucrates ductor*

Flanks boldly marked with five to seven dark bands. Caudal peduncle shows a small, fleshy keel. Swims freely in association with large sharks, fish, and rays. Also commonly seen under floating Gulfweed rafts. Body very rough and scaly to the touch. **Range:** Deep ocean waters worldwide. **Size:** To 27 in. (69 cm).

REMORA   *Remora remora*

Elongated body and flattened head with oval sucking disk. Lower jaw projects past upper jaw. Body color brown to dark brown. Attaches mainly to larger sharks, but may attach to any large ray, fish, or turtle. **Range:** Nova Scotia through tropics, usually in deeper waters. **Size:** To 30 in. (76 cm).

SHARKSUCKER   *Echeneis naucrates*

Elongated body very similar in shape and size to the Remora. Color varies, but alway has a wide, dark stripe down center of sides, bounded by thinner white stripes on top and bottom. Often swims free of hosts but prefers to attach itself to large fish, sharks, and turtles. Often caught by sportfishers in shallow coastal waters. **Range:** Nova Scotia through tropics. **Size:** To 36 in. (91 cm).

WHITEFIN SHARKSUCKER   *Echeneis neucratoides*

Very similar to the Sharksucker, but with extensive white areas in dorsal, anal, and tail fins. White stripes above and below dark central stripe very prominent. Often swims free of hosts but prefers to attach itself to large fish, sharks, and turtles. Much less common than the Sharksucker. **Range:** Nova Scotia through tropics. **Size:** To 30 in. (76 cm).

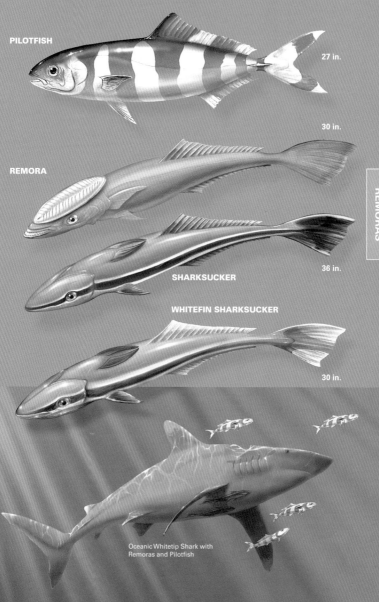

PILOTFISH

27 in.

30 in.

REMORA

SHARKSUCKER

36 in.

WHITEFIN SHARKSUCKER

30 in.

Oceanic Whitetip Shark with
Remoras and Pilotfish

SHARKS

LEMON SHARK
Negaprion brevirostris

Usually yellow-brown, but may be darker brown, with olive green flanks. Second dorsal fin almost as large as first. Tolerates brackish, low-oxygen waters of river mouths and other shallow areas. Also found in deeper coastal waters. **Range:** Uncommon in North Carolina, more frequent in south Florida. **Size:** To 5–8 ft. (1.5–2.4 m). **Red List – Near threatened**

NIGHT SHARK
Carcharhinus signatus

A medium-sized oceanic schooling shark that prefers deep waters, but enters coastal waters at night to feed. Slim, gray to gray-brown, long-snouted, with a low mid-dorsal ridge along the back. **Range:** Delaware to south Florida, Bahamas, and Cuba. Absent in Gulf of Mexico and Caribbean. **Size:** To 8–9 ft. (2.4–2.7 m). **Red List – Vulnerable**

ATLANTIC SHARPNOSE SHARK
Rhizoprionodon terraenovae

A small shark with a long, flattened snout, gray-blue to gray-brown. Adults often with small spots along flanks. Prefers inshore and coastal waters. **Range:** Common from South Carolina to south Florida and Gulf coasts to Texas, Bahamas, and Caribbean. Strays north to Cape Cod in small numbers. **Size:** To 2–3 ft. (61–91 cm).

CARIBBEAN REEF SHARK
Carcharhinus perezi

The most common reef and inshore island shark of the Caribbean and Bahamas. Less common in Florida and Gulf Coast waters. Similar to the Silky Shark, but much more stout-bodied, and Silkys prefer deep waters. Dorsal fin starts well after the trailing edge of the pectoral fin. **Range:** Florida and Gulf coasts, Bahamas, and West Indies. **Size:** To 5–8 ft. (1.5–2.4 m). **Red List – Near threatened**

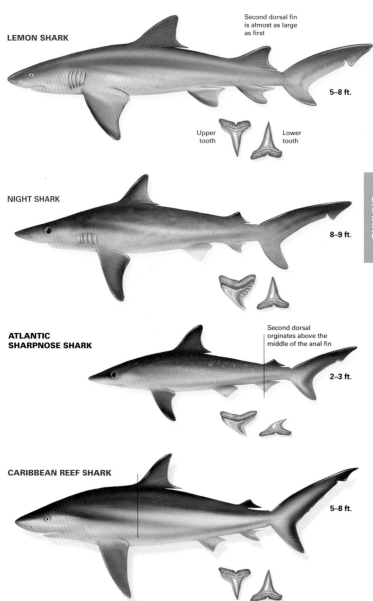

LEMON SHARK

Second dorsal fin is almost as large as first

5–8 ft.

Upper tooth   Lower tooth

NIGHT SHARK

8–9 ft.

ATLANTIC SHARPNOSE SHARK

Second dorsal orginates above the middle of the anal fin

2–3 ft.

CARIBBEAN REEF SHARK

5–8 ft.

SANDBAR SHARK (BROWN SHARK)   *Carcharhinus plumbeus*

Dark gray to brown and very pale below. Very high dorsal fin. The most common shark in shallower coastal waters. Enters bays with muddy bottoms. Migrates south in schools during late fall, when most bay sightings are made. **Range:** Massachusetts to Brazil. More common north of Cape Hatteras. **Size:** To 10 ft. (3 m). **Red List – Vulnerable**

DUSKY SHARK   *Carcharhinus obscurus*

Dark gray to brownish. Long, rounded snout. Distinct mid-dorsal ridge between small dorsal fins. Very large eyes. Mainly pelagic but occasionally wanders into coastal waters and river mouths. **Range:** Georges Bank to Brazil. **Size:** To 12 ft. (3.7 m). **Red List – Vulnerable**

BULL SHARK   *Carcharhinus leucas*

Gray to muddy brown-white below. Large-bodied, short snout. No mid-dorsal ridge. One of the most common large sharks. Very dangerous and known to attack humans. Common in coastal and offshore waters. **Range:** Massachusetts to Brazil. **Size:** To 12 ft. (3.7 m). **Red List – Near threatened**

SAND TIGER (SAND SHARK)   *Carcharias taurus*

Grayish brown to tan. Common in coastal waters. Often hunts in groups and at such shallow depths that its back is well exposed. When feeding at the surface, it often expels air through its gills, creating a sound like a whale spouting. **Range:** Throughout temperate Atlantic, Gulf, and Caribbean coasts. **Size:** To 10.5 ft. (3.2 m). **Red List – Vulnerable**

NURSE SHARK   *Ginglymostoma cirratum*

Rusty to yellowish brown. Note barbel on side of each nostril, used to locate shellfish and crustaceans. Rounded dorsal fin positioned far back over pelvic fin. Prefers coastal waters and bays. Likes reefs. **Range:** As far north as Rhode Island but more common south of North Carolina. **Size:** To 14 ft. (4.3 m). **Red List – Data deficient**

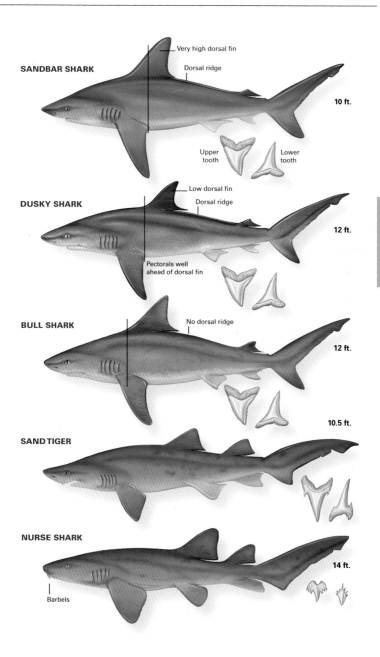

SANDBAR SHARK

Very high dorsal fin

Dorsal ridge

10 ft.

Upper tooth

Lower tooth

DUSKY SHARK

Low dorsal fin

Dorsal ridge

12 ft.

Pectorals well ahead of dorsal fin

BULL SHARK

No dorsal ridge

12 ft.

SAND TIGER

10.5 ft.

NURSE SHARK

14 ft.

Barbels

INSHORE SHARKS

75

THRESHER SHARKS, HAMMERHEADS

THRESHER SHARK
Alopias vulpinus

An extraordinarily long upper caudal fin is diagnostic of threshers and separates these large sharks from all others. Brownish to gray-brown. Normal eye size. **Range:** Gulf of St. Lawrence to Florida. Occurs off North Atlantic Coast mainly in summer. **Size:** To 20 ft. (6.1 m). **Red List – Data deficient**

BIGEYE THRESHER
Alopias superciliosus

Appearance like the Thresher Shark, but note its very large eyes, positioned high on the head to allow it to view upward. Back humped, dorsal fin set far back. **Range:** Throughout region north to New York. **Size:** To 18 ft. (5.5 m).

BONNETHEAD SHARK
Sphyrna tiburo

Grayish, with flattened head in form of shovel with eyes at edge of expanded portion. Abundant in bays and shallows into estuaries. **Range:** Entire Atlantic and Gulf Coasts. **Size:** To 6 ft. (1.8 m).

SMOOTH HAMMERHEAD
Sphyrna zygaena

Almost identical to the Scalloped Hammerhead, but without indentation in forehead and black tip on pectoral fin. **Range:** Entire Atlantic and Gulf Coasts. **Size:** To 13 ft. (4 m).

SCALLOPED HAMMERHEAD
Sphyrna lewini

Light brown or gray above, pale below. Note convex forehead with distinct indentation in front. Pectorals black on inside tips. Enters bays and shallow waters. **Range:** Entire Atlantic and Gulf Coasts. **Size:** To 10 ft. (3 m). **Red List – Data deficient**

GREAT HAMMERHEAD
Sphyrna mokarran

The largest and most pelagic and tropical of the hammerheads. Front edge of head slightly indented, giving it a very square appearance. Back edge of pelvic fin very curved. **Range:** North Carolina to Brazil; tropical waters worldwide. **Size:** To 20 ft. (6.1 m). **Red List – Data deficient**

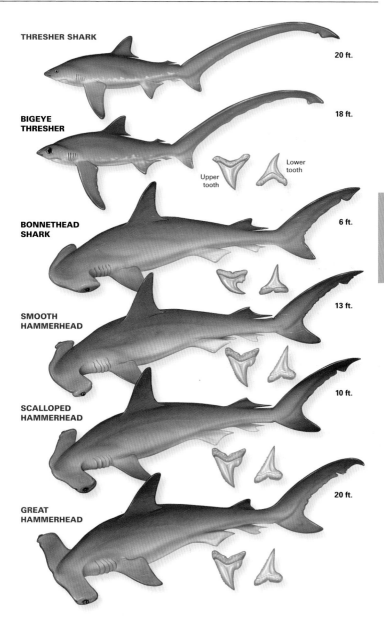

THRESHER SHARK

20 ft.

BIGEYE THRESHER

18 ft.

Upper tooth

Lower tooth

BONNETHEAD SHARK

6 ft.

SMOOTH HAMMERHEAD

13 ft.

SCALLOPED HAMMERHEAD

10 ft.

GREAT HAMMERHEAD

20 ft.

SHARKS

SIXGILL SHARK, DOGFISH

SIXGILL SHARK
Hexanchus griseus

Single dorsal fin, very long dorsal lobe of caudal fin, and six gill slits. Prefers deep warm waters, surfacing at night to feed. Color variable from tan to brown. Uncommon; habits poorly known. **Range:** North Carolina through Gulf of Mexico. **Size:** To 16 ft. (4.9 m). **Red List – Near threatened**

CHAIN CATSHARK (CHAIN DOGFISH)
Scyliorhinus retifer

Bold linear "chain" markings over a dusky brown back and lighter belly. A small, sluggish dogfish mostly found in deeper offshore waters. **Range:** Cape Cod south to Caribbean and Gulf of Mexico. **Size:** To 1.5 ft. (46 cm).

CUBAN DOGFISH
Squalus cubensis

Slim, gray or gray-brown, with distinct black patches at tips of dorsal fins. Otherwise very similar to the Spiny Dogfish, except for tropical distribution. **Range:** South Florida, Gulf Coast to Texas, and Cuba. Strays north to Cape Hatteras. **Size:** To 2–3.5 ft. (0.6–1.1 m).

SPINY DOGFISH
Squalus acanthias

Very small, brown to gray, with spotted flanks. Both dorsal fins show prominent spines at leading edges. Very common; often swims in large schools. Not fished commercially in our area, but popular in Europe as the main species served in "fish and chips." **Range:** Newfoundland to North Carolina and occasionally to Florida. **Size:** To 4 ft. (1.2 m).

SMOOTH DOGFISH
Mustelus canis

Very small, slender dogfish *without* spines along leading edges of either dorsal fin. Prominent spiracle opening behind eye. Enters bays in northern waters; prefers deeper, cooler water in southern parts of range. Very common; often caught by sportfishers. **Range:** Bay of Fundy to Uruguay. **Size:** To 5 ft. (1.5 m).

BLACK DOGFISH
Centroscyllium fabricii

Small, dark brown, with black belly and prominent spines forward of both dorsal fins. Second dorsal fin larger than first. Older specimens may be all black. Very common in northern parts of range, preferring deep, cold waters. **Range:** Southern Greenland to North Carolina. **Size:** To 3.5 ft. (1.1 m).

SIXGILL SHARK

Single dorsal fin

16 ft.

CHAIN CATSHARK

1.5 ft.

CUBAN DOGFISH

Black patches on both dorsal fins

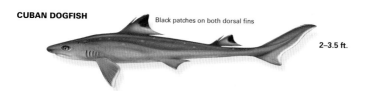

2–3.5 ft.

SPINY DOGFISH

Spotted flanks

Spines

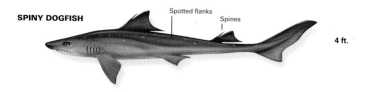

4 ft.

SMOOTH DOGFISH

No spines

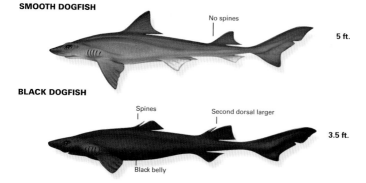

5 ft.

BLACK DOGFISH

Spines

Second dorsal larger

3.5 ft.

Black belly

WINTER SKATE
Leucoraja ocellata

Head and wings have rounded angles. Light brown with many small dark spots and usually one to four ocelli, or eye spots, on each wing. Typically found offshore at depths of 300 ft. (91 m) or more. **Range:** Newfoundland to North Carolina. **Size:** To 41 in. (102 cm).

ATLANTIC TORPEDO
Torpedo nobiliana

Rounded disc with a squared-off front edge. Typically dark gray-brown or violet-gray. Largest member of the electric ray family, and capable of generating 200-volt shocks if provoked. **Range:** Nova Scotia to North Carolina, straying south to Florida in deep waters far offshore. **Size:** To 6 ft. (1.8 m).

BARNDOOR SKATE
Raja laevis

Sharply angled wings, with finely pointed snout. Medium to dark brown, with many darker spots over the disc. **Range:** Nova Scotia to North Carolina, preferring deeper coastal waters near the southern edge of its range. **Size:** To 5 ft. (1.5 m), but typically 3–4 ft. (0.9–1.2 m).

ATLANTIC ANGEL SHARK
Squatina dumeril

Looks like a cross between a small shark and a skate. A row of toothlike denticles extends down the middle of the back forward of the first dorsal fin. Gray-brown to bluish gray, with a mouth at the leading edge of the head. **Range:** Cape Cod to Cape Hatteras, straying south to Florida in deep offshore waters. **Size:** To 5 ft. (1.5 m), but typically 3–4 ft. (0.9–1.2 m).

ATLANTIC GUITARFISH
Rhinobatos lentiginosus

Guitarfish have bodies like rays in front and like dogfish in the rear, with two large dorsal fins and a sharklike tail. Dark brown to gray with a fine spot pattern along flanks and wings. **Range:** North Carolina to Florida and Gulf Coast. **Size:** To 20–30 in. (61–76 cm).

CLEARNOSE SKATE
Raja eglanteria

Named for the translucent parts of the wings on either side of the pointed snout. Brown to gray-brown or violet-brown. Sharply angled wings. Usually marked with a pattern of irregularly shaped spots. **Range:** Massachusetts to Florida and Gulf Coast. **Size:** To 5 ft. (1.5 m), but averages 2–4 ft. (0.6–1.2 m).

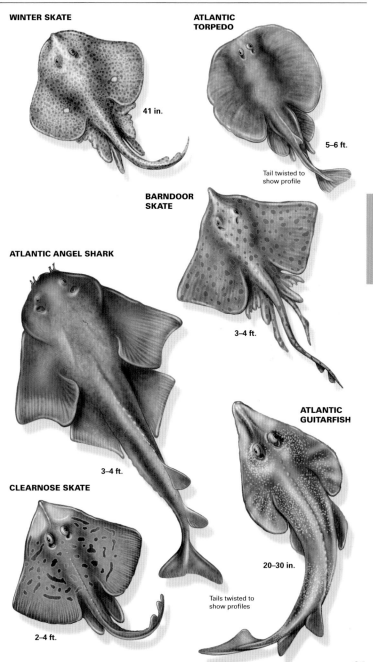

WINTER SKATE

41 in.

ATLANTIC TORPEDO

5–6 ft.

Tail twisted to show profile

BARNDOOR SKATE

3–4 ft.

ATLANTIC ANGEL SHARK

3–4 ft.

ATLANTIC GUITARFISH

20–30 in.

CLEARNOSE SKATE

2–4 ft.

Tails twisted to show profiles

STINGRAYS

BLUNTNOSE STINGRAY
Dasyatis say

Very rounded disc, with a row of very small spines down the center of the back. Dark brown to dark gray. **Range:** New Jersey to Florida and Gulf Coast, in shallow coastal waters. **Size:** To 3 ft. (91 cm) across the disc.

SOUTHERN STINGRAY
Dasyatis americana

Diamond-shaped disc, with sharp corners and a sharp snout. Uniformly brown or gray-brown, with a row of fine denticles down the center of the back. **Range:** New Jersey to Florida and Gulf Coast. **Size:** To 6 ft. (1.8 m) across the disc.

ATLANTIC STINGRAY
Dasyatis sabina

Disc with a sharp snout and very rounded wing corners. Uniform red-brown to dark brown. Common, particularly just offshore along the Gulf Coast. **Range:** New Jersey to Florida and Gulf Coast. Enters freshwater and larger rivers. **Size:** To 2 ft. (61 cm) across the disc.

ROUGHTAIL STINGRAY
Dasyatis centroura

Similar to the Southern Stingray, but with more rounded wing angles and with small spines irregularly scattered across the disc instead of a central row. The tail also has many small spines along the length. **Range:** Cape Cod to North Carolina, rarely to northern Florida. **Size:** To 7 ft. (2.1 m) across the disc, and up to 14 ft. (4.3 m) long.

PELAGIC STINGRAY
Dasyatis violacea

Leading edge of the disc is a broad smooth arc, with only a slight midline nose. Dark brown or dark gray. Row of spines along the midline of the disc. **Range:** Worldwide in deep offshore waters, but rare on Atlantic Coast south to Cape Hatteras. **Size:** To 3 ft. (91 cm) across the disc.

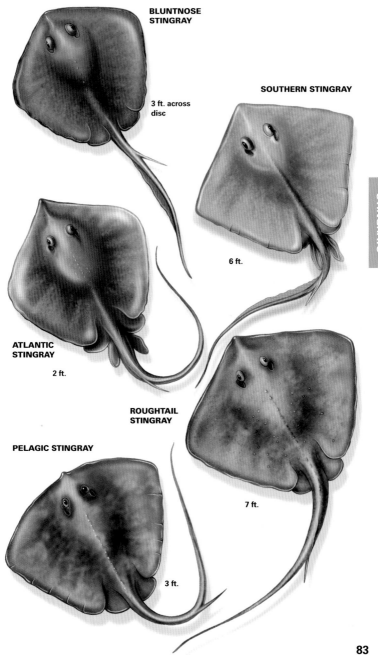

BLUNTNOSE STINGRAY

3 ft. across disc

SOUTHERN STINGRAY

6 ft.

ATLANTIC STINGRAY

2 ft.

ROUGHTAIL STINGRAY

7 ft.

PELAGIC STINGRAY

3 ft.

BUTTERFLY RAYS, EAGLE RAYS

SPINY BUTTERFLY RAY
Gymnura altavela

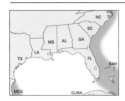

Very wide wings, making this ray wider than it is long. Dark gray or gray-brown, with a suble blotchy pattern across the disc. Uncommon throughout its range. **Range:** Cape Cod to Florida, largely absent along US Gulf Coast. **Size:** To 7 ft. (2.1 m) across the disc, but averages 4–6 ft. (1.2–1.8 m). **Red List – Vulnerable**

SMOOTH BUTTERFLY RAY
Gymnura micrura

Smaller than Spiny Butterfly Ray, with more prominent snout and pointed wing angles. Wider than it is long. Brown to gray-brown. **Range:** Maryland to southern Florida and Gulf of Mexico. **Size:** To 4 ft. (1.2 m) across disc.

BULLNOSE RAY
Myliobatis freminvillei

Sharply pointed wing angles, very prominent snout. Uniformly dark brown, with prominent dorsal fin at base of tail. **Range:** Cape Cod south to Florida, but absent from Gulf of Mexico. **Size:** To 3 ft. (91 cm) across the disc.

COWNOSE RAY
Rhinoptera bonasus

Sharply angled wingtips, massive snout that almost appears split with a midline groove. Forms large schools that mass offshore, particularly in autumn in Gulf of Mexico. **Range:** Cape Cod to Florida and Gulf of Mexico. **Size:** To 3 ft. (91 cm) across the disc. **Red List – Near threatened**

SOUTHERN EAGLE RAY
Myliobatis goodei

Similar to the Bullnose Ray, but with a much smaller dorsal fin set farther out on the tail. Wing angles are rounded, not in sharp points. Snout is less prominent than in the Bullnose Ray. **Range:** South Carolina to Florida; absent in Gulf of Mexico. **Size:** To 3 ft. (91 cm) across the disc.

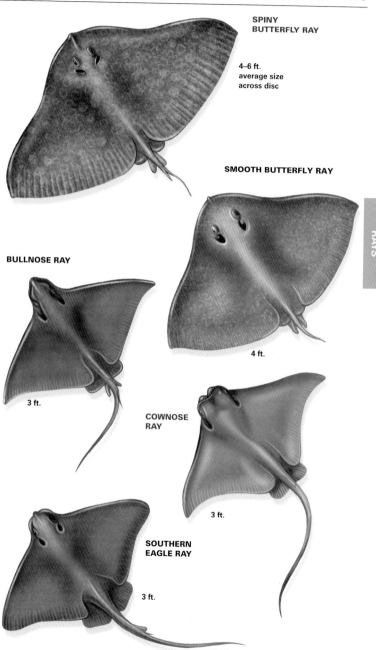

SPINY BUTTERFLY RAY

4–6 ft.
average size
across disc

SMOOTH BUTTERFLY RAY

BULLNOSE RAY

4 ft.

3 ft.

COWNOSE RAY

3 ft.

SOUTHERN EAGLE RAY

3 ft.

RAYS

Pelagic rays are more common well offshore in warmer waters, but they do enter coastal waters. Atlantic Mantas often cruise offshore channel areas and the outer reaches of reefs, feeding on plankton. Mantas and Spotted Eagle Rays are generally solitary, whereas Devil Rays often appear in schools. Mantas may breach spontaneously near boats, and all the larger rays may leap from the ocean surface if pursued.

ATLANTIC MANTA (DEVILFISH)   *Manta birostris*

The largest ray, can reach giant size. Disc black above, almost twice as wide as long, often showing gray or whitish patches near shoulders. Undersurface ranges from black or gray to almost white. Prominent cephalic fins, or "horns," flank terminal mouth (at leading edge of disc, not recessed as in Devil Ray). Often seen at or just below the surface. Harmless unless attacked or tangled in fishing or boating gear. **Range:** Cape Cod to Brazil. **Size:** Width to 22 ft. (6.7 m). **Red List – Near threatened**

DEVIL RAY   *Mobula hypostoma*

Very similar to a small Atlantic Manta. Disc black above, white or light gray below. Mouth subterminal rather than terminal as in the similar Manta. Tail length equals body length, proportionately much longer than the Manta's relatively short tail. Travels in schools. **Range:** Warm southern waters to Brazil; rarely as far north as New Jersey in summer. **Size:** Width to 4 ft. (1.2 m).

SPOTTED EAGLE RAY   *Myliobatis aquila*

Disc blue-gray, gray, or brown above, with lighter spots varying in size and shape. Head distinct from body. Tail very long and whiplike, with two spines at base. Favors shallow coastal waters; often seen in bays, coral reefs, inlets, and estuaries. **Range:** North to Cape Hatteras in summer; tropical waters worldwide. **Size:** Width to 8 ft. (2.4 m). **Red List – Near threatened**

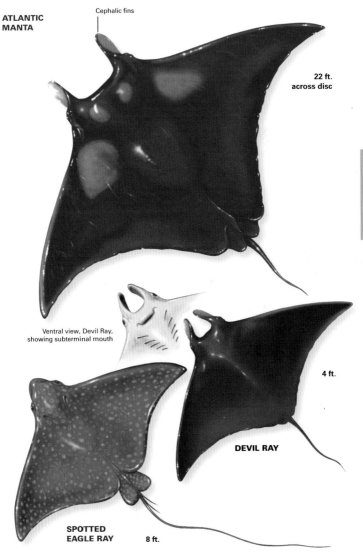

**ATLANTIC
MANTA**

Cephalic fins

**22 ft.
across disc**

Ventral view, Devil Ray,
showing subterminal mouth

4 ft.

DEVIL RAY

**SPOTTED
EAGLE RAY**   8 ft.

TARPON, BONEFISH

SHAFTED (LONGFIN) BONEFISH
Albula nemoptera

Similar in all respects to the much more common and larger Bonefish, but with an elongated last ray on the dorsal fin and a mouth that extends to below the eye. **Range:** Rare in south Florida and Florida Keys, more common in Caribbean. **Size:** To 20 in. (51 cm).

BONEFISH
Albula vulpes

Silver gray, with underslung mouth and dark green back ridge. Usually encountered in grass flats and sandy shallows, often near mangroves, particularly when the tide is rising. Retreats to deeper waters when not feeding. Prized in sportfishing for its fighting ability but rarely eaten. **Range:** Wanders north along Atlantic Coast, but rare north of Florida, Bahamas, and Caribbean. **Size:** To 40 in. (102 cm), but typically 20–30 in. (51–76 cm).

LADYFISH
Elops saurus

Small, sleek relative of the much larger Tarpon. Beautiful elongated, silvery fish, with dark back and subtle yellows and pinks along flanks. Darker fins and tail. Relatively large mouth extends to below the eye. Frequents coastal waters and shallows, particularly near mangroves. **Range:** North to Cape Cod, but rare north of Cape Hatteras, and common only in Florida and northern Gulf of Mexico. **Size:** To 36 in. (91 cm).

TARPON
Megalops atlanticus

Huge, metallic silver fish, prized in sportfishing for its fighting ability. Lower jaw projects to form a streamlined "snout." Fast, powerful swimmer, found in most coastal waters, from bays and inlets to deeper offshore reefs. **Range:** Wanders north in summer to at least Long Island, but increasingly uncommon north of Florida. **Size:** To 8 ft. (2.4 m), but typically 4–6 ft. (1.2–1.8 m).

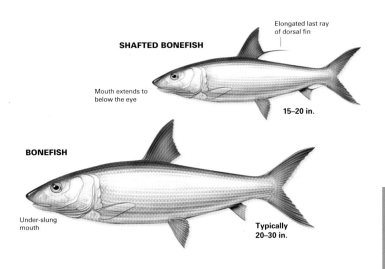

SHAFTED BONEFISH

Elongated last ray of dorsal fin

Mouth extends to below the eye

15–20 in.

BONEFISH

Under-slung mouth

Typically 20–30 in.

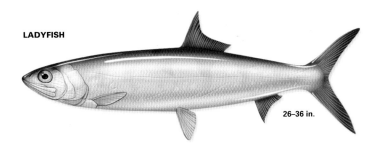

LADYFISH

26–36 in.

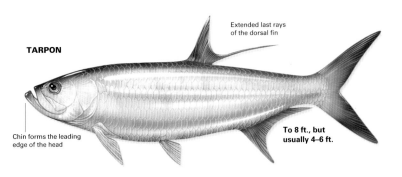

Extended last rays of the dorsal fin

TARPON

Chin forms the leading edge of the head

To 8 ft., but usually 4–6 ft.

GREEN MORAY

Gymnothorax funebris

The most common and familiar moray, frequently seen on coral reefs and in other shallow coastal waters where it can find its preferred shelter of rocks or small caves. Fierce-looking, but harmless unless provoked. **Range:** Wanders north to New Jersey, common in Florida, Bahamas, Gulf Coast, and Caribbean. **Size:** To 8 ft. (2.5 m), but usually much smaller.

SPOTTED MORAY

Gymnothorax moringa

Smaller moray of warm waters, cream-colored with heavy pattern of red-brown to brown spotting. Like all morays, favors small caves and crannies where it can shelter and ambush its prey. All morays can deliver a painful, infectious bite if handled or provoked, but they are normally not aggressive to humans. **Range:** Florida, Bahamas, Gulf Coast, and Caribbean, rare north of Florida waters. **Size:** To 36 in. (90 cm).

CHAIN MORAY

Echidna catenata

Small, beautiful moray, red-brown with crazed "chain" pattern of cream or yellow bands. Same habitat and behaviors as other morays, favoring sheltered caves in reefs and rocky areas. **Range:** Rare north of Cape Canaveral and southwest Florida, more common in Bahamas and Caribbean. **Size:** To 20 in. (50 cm).

CONGER EEL

Conger oceanicus

Large gray-brown to bluish gray eel common in Atlantic coastal waters. Dorsal fin begins above pectoral fin. Favors rocky coastal waters, reefs, and piers where it can shelter. **Range:** Cape Cod to mid-Florida and northern Gulf of Mexico. **Size:** To 7.5 ft. (2.3 m).

AMERICAN EEL

Anguilla rostrata

Greenish brown to red-brown back, grading to yellow-brown below. Dorsal fin begins well to the rear of the pectoral fin, about halfway between pectoral fin and anus. **Range:** Common all along the Atlantic and Gulf Coasts, Bahamas, and Caribbean. **Size:** To 5 ft. (1.5 m), but usually smaller.

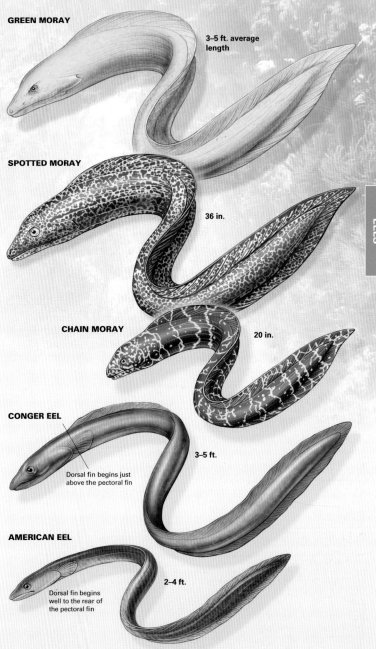

GREEN MORAY

3–5 ft. average length

SPOTTED MORAY

36 in.

CHAIN MORAY

20 in.

CONGER EEL

Dorsal fin begins just above the pectoral fin

3–5 ft.

AMERICAN EEL

Dorsal fin begins well to the rear of the pectoral fin

2–4 ft.

HERRING, SHAD, MENHADEN

A large family of some 27 species that have long been staples of the fishing industry. As with many ocean fish, overfishing of these species has contributed to declining populations.

AMERICAN SHAD
Alosa sapidissima

Extremely laterally compressed with dark back and silvery flanks showing a row of spots on upper portion. Known for famous "shad runs" up such large rivers as the St. Johns River in central Florida. **Range:** New Brunswick to Florida. **Size:** To 30 in. (76 cm).

ATLANTIC MENHADEN (MOSSBUNKER)
Brevoortia tyrannus

Small version of the American Shad, with distinctly larger head in proportion to body. Occurs in large schools that support an East Coast fishery producing fish oil for the market. **Range:** New Brunswick to Florida. **Size:** To 18 in. (46 cm).

HICKORY SHAD
Alosa mediocris

Nearly identical to American Shad, but mouth turns distinctly upward and side spotting starts right next to gill cover. Dark blue back. During "shad runs" up rivers can become infected with water molds that produce large lesions. **Range:** Maine to Florida. **Size:** To 26 in. (66 cm).

ATLANTIC HERRING
Clupea harengus

 Elongate silver fish, less laterally flattened than most in family. Occurs in massive schools that frequent the open ocean surface. Caught by large fisheries, it is also taken by offshore fishing boats and provides food for whales and larger sport fish. **Range:** Greenland to North Carolina. **Size:** To 18 in. (46 cm).

BLUEBACK HERRING
Alosa aestivalis

Very similar to the Alewife in color patterns and behavior. Mixes with shad and Alewife during runs. Bluebacks have smaller eyes and more blue-green back but can be hard to separate from similar herrings. **Range:** Nova Scotia to Florida. **Size:** To 14 in. (36 cm).

ALEWIFE
Alosa pseudoharengus

 Strongly dorsally compressed with a blue-black back and silvery flanks. Moves in incredible numbers known as "Alewife runs" from coastal and estuarine areas to inland breeding sites. Important food species for Striped Bass, Bluefish, and other sport fish. **Range:** Newfoundland to South Carolina. **Size:** To 15 in. (38 cm).

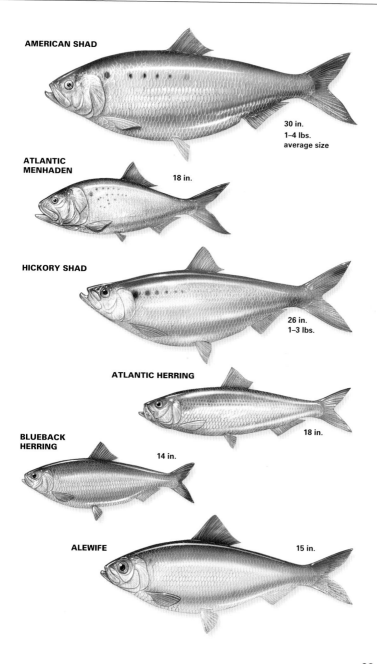

AMERICAN SHAD

30 in.
1–4 lbs.
average size

ATLANTIC MENHADEN

18 in.

HICKORY SHAD

26 in.
1–3 lbs.

ATLANTIC HERRING

18 in.

BLUEBACK HERRING

14 in.

ALEWIFE

15 in.

HERRING

93

COD

ATLANTIC COD
Gadus morhua

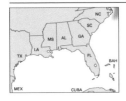

Bottom-dwelling fish of deep waters. Brownish to olive green, heavily spotted. Note chin barbel. The population is overfished, and cod are declining rapidly all along the East Coast. Weight now averages less than 20 lbs. (9 kg), although past specimens sometimes reached 200 lbs. (91 kg). **Range:** Greenland to North Carolina. **Size:** Typically 12–24 in. (30–61 cm). **Red List – Vulnerable**

POLLOCK
Pollachius virens

Slimmer than a cod, this greenish olive relative with protruding lower jaw lacks a noticeable chin barbel. Fast swimmers, Pollock will school to capture other fish and will come to the surface to feed. **Range:** Newfoundland to North Carolina, with greatest concentrations in Gulf of Maine. **Size:** To 35 in. (89 cm).

HADDOCK
Melanogrammus aeglefinus

Excellent food fish of cool, deep waters. Note dark spot above pectoral fin. **Range:** Newfoundland to North Carolina; more common in northern range. **Size:** Typically to 24 in. (61 cm). **Red List – Vulnerable**

SILVER HAKE
Merluccius bilinearis

Trim silver fish with protruding lower jaw and lacking typical cod chin barbel. Rear dorsal and anal fins almost mirror each other in shape. Travels in small groups, attacking prey fish in a feeding frenzy. Populations are split into northern and southern stocks. **Range:** Newfoundland to North Carolina. **Size:** To 30 in. (76 cm).

WHITE HAKE
Urophycis tenuis

Elongate, whiskerlike ventral fins separate this fish from others in group. These long fins are used for hunting in the dark in deep, cold waters for crabs and mollusks. **Range:** Newfoundland to Cape Hatteras. **Size:** To 48 in. (1.2 m).

ATLANTIC COD

Heavily spotted body

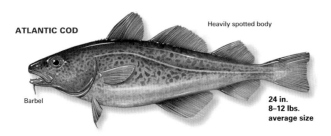

Barbel

**24 in.
8–12 lbs.
average size**

POLLOCK

Projecting
lower jaw

Barbel

**35 in.
2–6 lbs.**

HADDOCK

Dark spot
above
pectoral fin

Barbel small
or absent

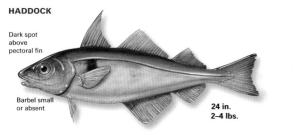

**24 in.
2–4 lbs.**

SILVER HAKE

Projecting lower
jaw; no barbel
under jaw

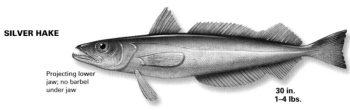

**30 in.
1–4 lbs.**

WHITE HAKE

**48 in.
2–6 lbs.**

COD

HALFBEAKS, TRUMPETFISH, SLIM-BODIED FISHES

See the next page spread for illustrations of these species.

GULF PIPEFISH
Syngnathus scovelli

Pipefish are elongated relatives of seahorses and share with them ringed ridges along the length of the body. Light brown to brown. Common, usually close to shore in bays and inlets, grass flats, sandbanks, and river mouths. **Range:** Florida and Gulf Coasts, rare north of Florida. **Size:** To 7 in. (18 cm).

DUSKY PIPEFISH
Syngnathus floridae

Larger, darker pipefish with less prominent body rings than in the Gulf Pipefish. Very common in grass flats, shallow banks, and bays. **Range:** Chesapeake Bay south to Florida, Gulf Coast, Bahamas, and Caribbean. **Size:** To 10 in. (25 cm).

SARGASSSUM PIPEFISH
Syngnathus pelagicus

Small, highly camouflaged pipefish that normally lives well offshore in mats of Gulf Weed. Colors vary but generally match surrounding *Sargassum* browns and yellows. **Range:** Maine to Florida, Bahamas, and Gulf of Mexico. **Size:** To 8 in. (20 cm).

AGUJON
Tylosurus acus

One of the larger needlefish, with beak relatively short in proportion to overall head size. Color usually browns and mustard yellows. Upper jaw slightly shorter than lower jaw. **Range:** Massachusetts to Brazil, Gulf of Mexico, but common mainly in tropical waters. **Size:** To 36 in. (91 cm).

FLAT NEEDLEFISH
Ablennes hians

A larger needlefish, body very flat from side to side. Silvery blue above grading to silver-white below, with diffuse vertical bars along flanks. Dark spot at rear edge of dorsal fin. **Range:** Chesapeake Bay to Florida, Gulf of Mexico, and Caribbean. **Size:** To 43 in. (109 cm).

BLUESPOTTED CORONETFISH
Fistularia tabacaria

Unusual elongated body form, usually bright to dark green or green-blue, with many bright blue irregularly shaped spots along flanks. Long caudal filament extends from tail fin. Common over reefs and in shallow banks in Bahamas and Caribbean, less common in Florida waters. **Range:** Cape Cod south to Brazil, southeastern Gulf of Mexico, Bahamas, and Caribbean. **Size:** To 6 ft. (1.8 m).

BALLYHOO
Hemiramphus brasiliensis

These warm-water halfbeaks (narrow-bodied fishes with elongated jaws) are related to flyingfish. Schools in large numbers near the water's surface and serves as food for such larger game fish as sailfish, billfish, and mackerel. Harvested mainly as baitfish for offshore fishing. **Range:** New York to Brazil, throughout Gulf of Mexico. **Size:** To 8 in. (21 cm).

BALAO
Hemiramphus balao

A Ballyhoo look-alike, and often called "Ballyhoo," especially in southeastern baitfish markets. Body silver, not greenish as in the Ballyhoo. Schools in large numbers near the water's surface, attracting the attention of game fish and seabirds, its principal predators. **Range:** New York to Brazil, northern Gulf. **Size:** To 9 in. (23 cm).

AMERICAN HALFBEAK
Hyporhamphus meeki

Note long, bright orange-red lower jaw. Favorite food of Dolphin, billfish, and mackerel. Large schools feed near the water's surface and are often seen skittering away from approaching boats. **Range:** Maine to Argentina, throughout Gulf of Mexico. **Size:** To 10.5 in. (27 cm).

ATLANTIC NEEDLEFISH
Strongylura marina

Active mainly at night, coming readily to night lights on bait boats to hunt small fish, its principal prey. Easily separated from other halfbeaks by brown back and silvery belly. **Range:** Maine to Brazil, throughout Gulf of Mexico and Caribbean. **Size:** To 25 in. (64 cm).

TRUMPETFISH
Aulostomus maculatus

Odd, elongated body shape, usually brown to red-brown, with dark spots along the back and banded flanks. It has an unusual habit of hanging above the reef head-down, almost vertical. **Range:** Southern Florida to Argentina, Bahamas, Caribbean, southeastern Gulf of Mexico. **Size:** To 36 in. (91 cm).

HOUNDFISH
Tylosurus crocodilus

The largest needlefish, recognizable by its great size alone. Dark blue or blue-gray above grading to silver or cream belly, and striped along flanks, particularly toward tail. Lower jaw longer than upper jaw. **Size:** To 5 ft. (1.5 m), but more often averages 3 ft. (91 cm).

HALFBEAKS, TRUMPETFISH, SLIM-BODIED FISHES

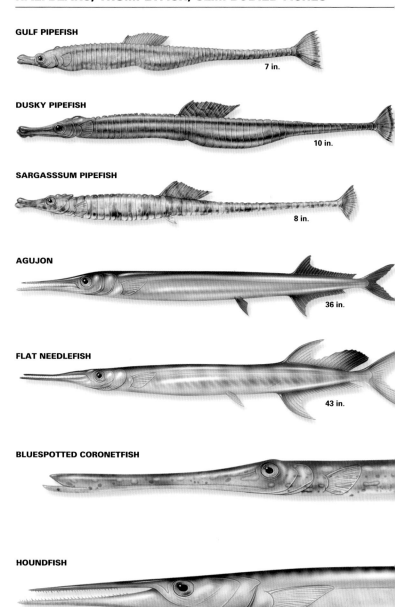

GULF PIPEFISH

7 in.

DUSKY PIPEFISH

10 in.

SARGASSSUM PIPEFISH

8 in.

AGUJON

36 in.

FLAT NEEDLEFISH

43 in.

BLUESPOTTED CORONETFISH

HOUNDFISH

BALLYHOO

8 in.

BALAO

9 in.

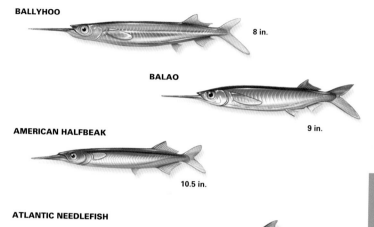

AMERICAN HALFBEAK

10.5 in.

ATLANTIC NEEDLEFISH

25 in.

TRUMPETFISH

36 in.

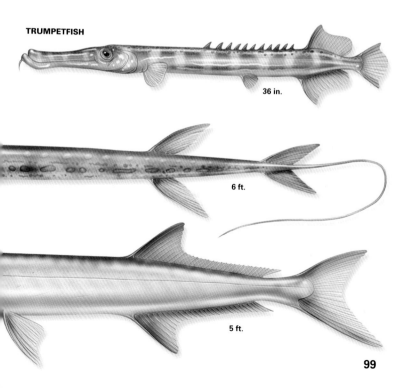

6 ft.

5 ft.

FLYINGFISH

This group favors tropical offshore waters. The species here frequent the Gulf Stream, and a few can be found in open temperate oceans. Flyingfish sometimes enter inshore waters after storms but are generally seen in warm offshore waters.

OCEANIC TWO-WING FLYINGFISH · *Exocoetus obtusirostris*

Very long pectoral fins, blunt head. Dorsal fin small and unmarked. The only two-winged flyingfish to reach our area. **Range:** New Jersey south, far offshore or in Gulf Stream. **Size:** To 10 in. (25 cm).

MARGINED FLYINGFISH · *Cypselurus cyanopterus*

Pectoral fins and caudal fin uniformly dusky. Large dark patch along upper edge of dorsal fin. **Range:** New Jersey south, far offshore or in Gulf Stream. **Size:** To 18 in. (46 cm).

BANDWING FLYINGFISH · *Cheilopogon exsiliens*

Spectacular dark blue pectoral fins with narrow, clear crossing bands. Caudal fin clear on top and dusky on lower portion. Common offshore. **Range:** Cape Cod south, far offshore or in Gulf Stream. **Size:** To 12 in. (30 cm).

ATLANTIC FLYINGFISH · *Cheilopogon melanurus*

Black pectoral fins with clear central area. Dusky caudal fin. The most common flyingfish of Gulf Stream waters. Often seen skittering away from the bow of boats in deep offshore waters. Very common in southern waters off Florida; unusual in northern waters. **Range:** Cape Cod south, far offshore or in Gulf Stream. **Size:** To 16 in. (41 cm).

SPOTFIN FLYINGFISH · *Cheilopogon furcatus*

Striking black pectoral fins with wide, clear crossing bands. Pelvic fins with dark tips and clear caudals. Commonly enters bays and shallow coastal waters in Florida and Gulf of Mexico. **Range:** Cape Cod south, far offshore or in Gulf Stream. **Size:** To 14 in. (36 cm).

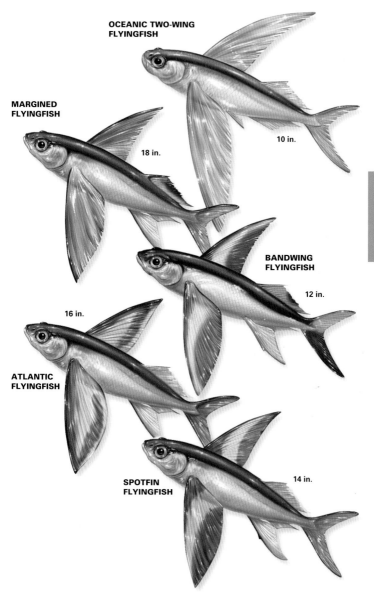

OCEANIC TWO-WING
FLYINGFISH

MARGINED
FLYINGFISH

18 in.

10 in.

BANDWING
FLYINGFISH

12 in.

16 in.

ATLANTIC
FLYINGFISH

SPOTFIN
FLYINGFISH

14 in.

FLYINGFISH

GROUPERS

RED HIND
Epinephelus guttatus

Densely covered with red spotting on cream base. Also called the Strawberry Grouper. Rear edge of dorsal fin and edges of caudal and anal fins black or dark brown. In shallow reefs and shoal areas. **Range:** Mainly off Florida and Bermuda, but some wander a bit farther north. **Size:** To 24 in. (46 cm).

SPECKLED HIND
Epinephelus drummondhayi

Another fish sometimes called the Strawberry Grouper. Dark red-gray to light red-brown, completely covered, including fins, with small whitish dots. Uncommon in deep water around ledges and outcrops. **Range:** Cape Hatteras to southern Florida, Gulf of Mexico, and Caribbean. **Size:** To 18 in. (46 cm). **Red List – Critically endangered**

RED GROUPER
Epinephelus morio

Rusty red with spotting and large dark or light blotches on body. Unmarked tail fin. Found inshore to offshore from shallow reefs to deeper waters. One of the few groupers that will fight when hooked. Formerly one of the most common groupers, it is fished commercially, and thousands of tons are caught each year off Florida and Mexico. **Range:** Massachusetts to Florida, Gulf of Mexico, and Caribbean. **Size:** To 28 in. (71 cm). **Red List – Near threatened**

NASSAU GROUPER
Epinephelus striatus

Large, pinkish brown. Five wide side bars darker in color run from above eye to base of tail, where a dark spot is found on top of the peduncle. In shallow coral reefs to depths of 70 ft. (21.3 m). Illegal to take in US waters. **Range:** North Carolina south to Brazil, southern Gulf of Mexico. **Size:** To 4 ft. (1.2 m). **Red List – Endangered**

RED HIND

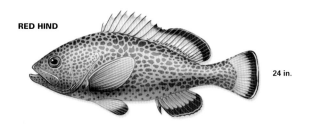

24 in.

SPECKLED HIND

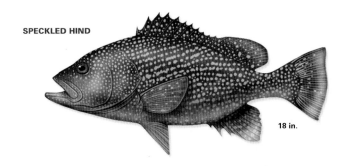

18 in.

RED GROUPER

28 in.

NASSAU GROUPER

4 ft.

GROUPERS

SCAMP

Mycteroperca phenax

Chocolate or tan, heavily dotted with dots that often fuse to form lines. Long rays of squared-off tail give a broomlike appearance, hence the alternate name Broom-Tailed Grouper. Usually found well offshore over deep banks. The top grouper for commercial fishing. **Range:** Cape Hatteras to Florida and Gulf of Mexico, wandering as far as Massachusetts. **Size:** To 30 in. (76 cm).

SAND PERCH

Diplectrum formosum

Handsome, slender, with horizontal blue and orange lines on a tan-yellow base. Blue lines on cheek. In fairly shallow water over sand and mud bottoms, but also near reefs. **Range:** Virginia to Florida and throughout Gulf of Mexico. **Size:** To 12 in. (30 cm).

ROCK HIND

Epinephelus adscensionis

Stocky, olive yellow with reddish brown spotting covering body and fins. Rear of dorsal fin lobed and dark. Anal fin dark edged. Favors shallow water over reefs and rocky bottoms and can tolerate rough waters inshore. **Range:** Bermuda to south Florida and Gulf Coast. **Size:** To 24 in. (61 cm).

SNOWY GROUPER

Epinephelus niveatus

Large, deep bodied, reddish tan, dotted with large pinkish to white dots. Bright white spots gives this fish its common name. Younger fish have a distinct black saddle to the top of the anal fin peduncle. Favors deeper waters over rocky bottoms. Younger fish occur in shallower coastal waters. **Range:** Cape Cod south to Caribbean and Gulf of Mexico. **Size:** To 4 ft. (1.2 m).

YELLOWEDGE GROUPER

Epinephelus flavolimbatus

Mottled light brown with distinct yellow edging trimmed with black on dorsal, pectoral, and anal fins. Prefers waters over coral reefs. Stressed by commercial fishing on its spawning grounds. **Range:** North Carolina south through Florida and Gulf of Mexico to Brazil. **Size:** To 24 in. (61 cm). **Red List – Vulnerable**

SCAMP

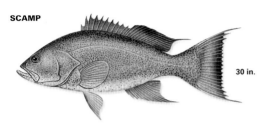

30 in.

SAND PERCH

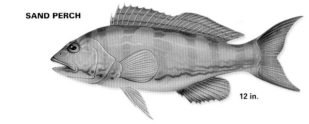

12 in.

ROCK HIND

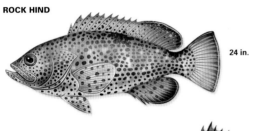

24 in.

SNOWY GROUPER

4 ft.

YELLOWEDGE GROUPER

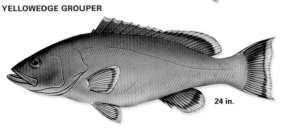

24 in.

GAG GROUPER
Mycteroperca microlepis

Gray to brown with bold patterns of brown often edged in blue. Pattern with thinner lines than in the similar Black Grouper. Shallow inshore waters to offshore over rocks and reefs. Illegal to take in US Gulf waters due to overfishing. **Range:** Cape Hatteras to Florida and Gulf of Mexico. **Size:** To 57 in. (1.4 m).

TRIPLETAIL
Lobotes surinamensis

The only species in this family with looks similar to the groupers. Large-bodied, with large, round lobes on dorsal and anal fins and a rounded caudal fin that gives the appearance of three tails. Favors inshore bays and estuaries often near buoys, especially in flotsam mats of algae. **Range:** Cape Cod south, most common south of Cape Hatteras and in Gulf of Mexico. **Size:** To 3 ft. (91 cm).

BLACK GROUPER
Mycteroperca bonaci

Elongate, less stocky than many groupers. Light brown with patches, often square in shape, all over body and fins. Posterior fins have black borders. Favors offshore reefs, rocky banks, and old wrecks. **Range:** Cape Hatteras to Florida and southern Gulf of Mexico, northern Gulf in deep waters. **Size:** To 4 ft. (1.2 m).

WARSAW GROUPER
Epinephelus nigritus

Dark reddish brown, lacking any significant markings. May almost be black above. Prefers deep waters, often around oil rig bases. **Range:** Cape Cod south to Florida and Gulf of Mexico. **Size:** To 5 ft. (1.5 m). **Red List – Critically endangered**

GOLIATH GROUPER
Epinephelus itajara

Huge. Young have brilliant rusty bands on a brownish to gray base. Young favor mangrove areas. Adults prefer bridge pilings, piers and docks that extend into deep water, and offshore oil rig bases. Also found in reefs and underwater rocky ledges. **Range:** Florida and Gulf of Mexico to Caribbean. **Size:** To 8 ft. (2.4 m). **Red List – Critically endangered**

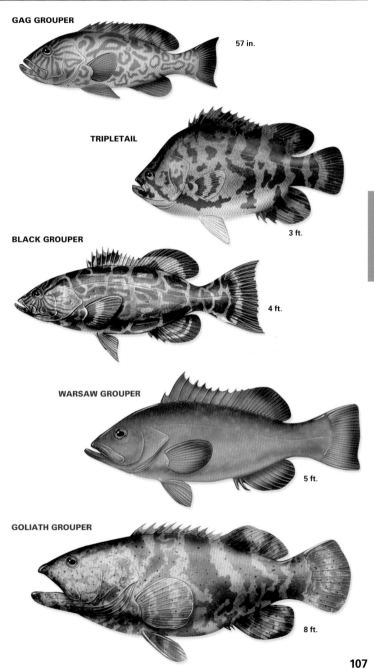

GAG GROUPER

57 in.

TRIPLETAIL

3 ft.

BLACK GROUPER

4 ft.

WARSAW GROUPER

5 ft.

GOLIATH GROUPER

8 ft.

LARGE
GROUPERS

GAME FISH

Most of these popular game fish are medium-sized predators that frequent both inshore waters and the deeper waters of the continental shelf. Populations of these species fluctuate, but most are considered to be in a long-term decline due to overfishing, pollution, and habitat loss.

GREAT BARRACUDA
Sphyraena borealis

Large and slim. Overshot lower jaw reveals numerous sharp teeth. Large, triangular tail. Can be aggressive toward divers, but rarely bites unless deliberately provoked. Flesh edible but may cause ciguatera poisoning. **Range:** Massachusetts south; prefers warm waters. **Weight:** To over 100 lbs. (45 kg). **Size:** To 6 ft. (1.8 m).

COBIA
Rachycentron canadum

Large, brownish, with dark side stripes and pale underparts. Lower jaw protrudes like a barracuda's, but teeth are smaller. Spines form first part of dorsal fin. Hunts near floating matter and buoys. **Range:** Massachusetts to Florida, Gulf of Mexico, and Caribbean. **Weight:** To 150 lbs. (68 kg). **Size:** To 6 ft. (1.8 m).

BLUEFISH
Pomatomus saltatrix

Blue above, silver below. Jaw has numerous very sharp teeth. Hunts in schools; voracious in feeding. Will drive fish into shallow areas and attack. Bluefish have even bitten bathers when in feeding frenzies. Very important game fish. **Range:** Nova Scotia to Argentina. **Weight:** To 27 lbs. (12 kg). **Size:** To 45 in. (1.1 m).

STRIPED BASS
Morone saxatilis

Large, swift, silvery, with seven or eight distinct dark side stripes. Very dependent on water temperature for movement in- and offshore. Important game and food fish. **Range:** Gulf of St. Lawrence to Florida and eastern Gulf Coast. **Weight:** Rarely above 45 lbs. (20 kg) but ranges to 125 lbs. (57 kg). **Size:** To 6 ft. (1.8 m) but typically 2–3 ft. (61–91 cm).

GREAT BARRACUDA

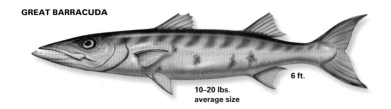

6 ft.

10–20 lbs.
average size

COBIA

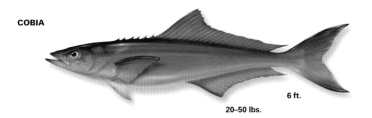

6 ft.

20–50 lbs.

BLUEFISH

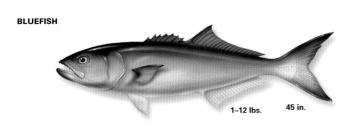

1–12 lbs.   45 in.

STRIPED BASS

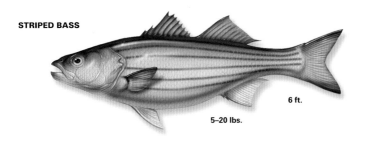

6 ft.

5–20 lbs.

POMPANOS

Pompanos are swift, nervous inshore predators. They make delicious eating but are challenging to hook.

PERMIT
Trachinotus falcatus

Large, silvery, disk-shaped fish. Note swept-back dorsal and anal fins. Dark blotch near pectoral fin is usually distinctive. **Range:** Massachusetts to Brazil. **Size:** To 45 in. (1.1 m).

FLORIDA POMPANO
Trachinotus carolinus

Very similar to the Permit but lacks dark blotch on flank. Back is often blue-black. Common in coastal bays and estuaries. Important food fish. **Range:** Massachusetts to Brazil. **Size:** To 25 in. (64 cm).

PALOMETA
Trachinotus goodei

Trim oval body with distinct side barring. Silvery blue with darker back. Elongate dorsal and anal fins with soft texture, dark on leading edges. Lower edge of tail fin also distinctly dark. Occurs in large schools hunting smaller fish. More common in subtropical waters. **Range:** Mainly Cape Hatteras to Caribbean, Gulf of Mexico. **Size:** To 13 in. (33 cm).

LOOKDOWN
Selene vomer

Nearly vertical profile of face, side stripes, and elongate anterior lobes of dorsal and anal fins make this very recognizable. Favors inshore shallows over sandy areas and around pilings, often in schools. **Range:** Mainly off southern East Coast but ranges north to Maine. Throughout Gulf of Mexico. **Size:** To 15 in. (38 cm).

AFRICAN POMPANO
Alectis ciliaris

Rays of dorsal and anal fins are reduced to distinctive threadlike projections. Young specimens, called Threadfish, have fin projections much longer than the body. Note very steep forehead angle. **Range:** Massachusetts to Brazil. **Size:** To 45 in. (1.1 m).

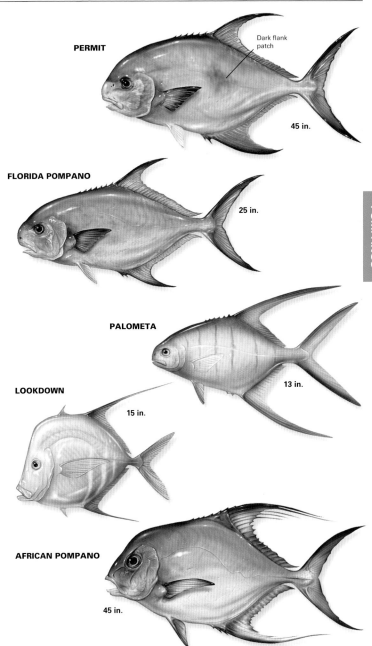

PERMIT

Dark flank patch

45 in.

FLORIDA POMPANO

25 in.

PALOMETA

13 in.

LOOKDOWN

15 in.

AFRICAN POMPANO

45 in.

111

JACKS

Jacks are voracious offshore game fish often found near reefs, oil platforms, and wrecks in deeper waters.

BLUE RUNNER
Caranx crysos

Deep "saddle" in dorsal fin. Back bluish green to olive green. Pale wavy lines cover flanks. Distinct black spot on rear of gill cover. Young favor offshore areas and exploit inshore habitats as they age. Coastal creeks, inlets, and bays to well offshore. **Range:** Nova Scotia to Florida and south, Gulf of Mexico. **Size:** To 12–14 in. (30–36 cm).

BAR JACK
Caranx ruber

Elongate, compressed, sliver to silver-blue, with distinct bar running from base of dorsal fin to lower lobe of caudal fin. Found in shallow waters inshore and over reefs. **Range:** New Jersey south, but more common from Cape Hatteras south to Venezuela, including Gulf of Mexico. **Size:** To 22 in. (56 cm).

LEATHERJACK (LEATHERJACKET)
Oligoplites saurus

Sleek, blue-green, with sharp spines on forward dorsal fin. Also beware of sharp anal fin spines. Bright yellow tail contrasts with silver-blue body. Favors inland waters into bays and estuaries. **Range:** North as far as Maine, but more common from Cape Hatteras south to Florida and Gulf of Mexico. **Size:** To 12 in. (30 cm).

BANDED RUDDERFISH
Seriola zonata

Small member of the amberjack family with distinct dark line running through eye to fore part of dorsal fin. Flanks marked by six distinct lines that fade with age. Found inshore in bays, channels, and often near marker buoys. **Range:** Cape Cod to Florida and Gulf of Mexico. **Size:** Usually 1 ft. (30 cm) but can reach 2 ft. (61 cm).

PILOTFISH
Naucrates ductor

Similar to the Banded Rudderfish but more slender; lacks black line running through eye, banding more distinct. Found offshore in company of sharks, rays, larger fish, and turtles, feeding on scraps. **Range:** East Coast and eastern Gulf of Mexico. **Size:** To 2 ft. (61 cm).

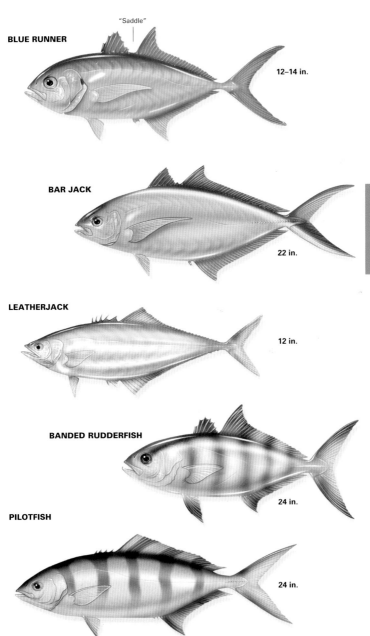

"Saddle"

BLUE RUNNER

12–14 in.

BAR JACK

22 in.

LEATHERJACK

12 in.

BANDED RUDDERFISH

24 in.

PILOTFISH

24 in.

JACKS

JACKS

ALMACO JACK
Seriola rivoliana

Dark line through eye extends onto back. Deep, compressed body, tapered head, and sickle-shaped dorsal and anal fins allow separation from the similar Amberjack. Found in inshore and offshore reefs. Attracted to wrecks. **Range:** Cape Hatteras to Florida and Gulf of Mexico. **Size:** To 32 in. (80 cm).

HORSE-EYE JACK
Caranx latus

Silvery blue with blunt forehead. Dorsal fin black-tipped, sometimes with black spot on gill cover. Found in shallow waters around sandy-edged islands usually in small schools. Also occurs offshore. **Range:** New Jersey south to Argentina including Gulf of Mexico. **Size:** To 30 in. (75 cm).

YELLOW JACK
Caranx bartholomaei

Beautiful, with light aqua green of flanks fading into silver-yellow. Light wavy bars mark flanks. Fins mainly yellow. Young fish stay inshore over sandy bottoms. Larger fish occur over reefs and cold water drop-offs. **Range:** Cape Hatteras to Florida and eastern Gulf of Mexico. Occasionally wanders northward. **Size:** To 39 in. (1 m).

CREVALLE JACK
Caranx hippos

Large, stocky, with blunt forehead. Beautiful blue above fades to silvery underparts. Black spot on gill cover. Wide ranging, from far offshore and deep waters to inshore and bays and even into river mouths at times. **Range:** Widespread from Nova Scotia to Uruguay and Gulf of Mexico. **Size:** To 5 ft. (1.5 m).

Yellow Jacks. Jacks are swift predators that move in schools across open water.

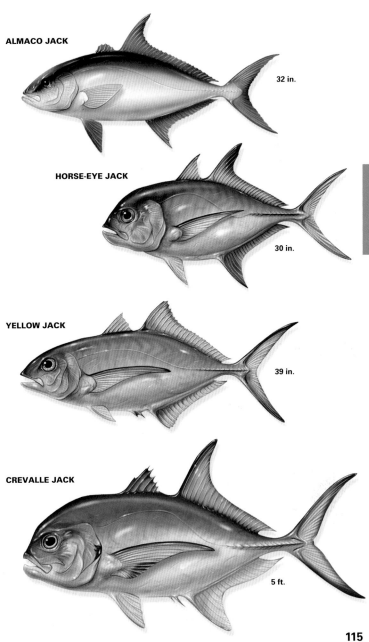

ALMACO JACK

32 in.

HORSE-EYE JACK

30 in.

YELLOW JACK

39 in.

CREVALLE JACK

5 ft.

JACKS

RAINBOW RUNNER
Elagatis bipinnulata

Sleek, elongate. Beautifully colored with flashing blues grading to yellow flanks that form a stripe contrasting with white below. Small finlets on top and bottom of anal peduncle. Often found near the surface over deepwater reefs. **Range:** Massachusetts to Brazil and Gulf of Mexico. **Size:** To 4 ft. (1.2 m).

BLUNTNOSE JACK
Hemicaranx amblyrhynchus

A deep bodied, pale yellowish white fish with yellowish dorsal and anal fins. Young develop inshore and show distinct side banding that is seen in many young jacks. In small schools in coastal waters and bays into estuaries. **Range:** North Carolina to Florida, eastern Caribbean, and northern Gulf of Mexico. **Size:** To 20 in. (50 cm).

LESSER AMBERJACK
Seriola fasciata

Similar to the young Greater Amberjack, but note that dark line through eye stops well short of dorsal fin. Dark back shades to yellow stripe on flanks. Often hunts edges of drifting algal mats. Found in both deep and shallow waters. **Range:** Cape Cod to Florida and eastern Gulf of Mexico. **Size:** To 12 in. (30 cm).

GREATER AMBERJACK
Seriola dumeril

Elongate, large, yellow-brown above fading to white below. Flanks often amber-colored, hence the name. Note that dark band through eye extends to base of dorsal fin. No finlets on peduncle. Often found in large schools in open waters and near reefs in both shallow and deep waters. **Range:** Cape Cod to Brazil and Gulf of Mexico. **Size:** To 5 ft. (1.5 m).

RAINBOW RUNNER

4 ft.

BLUNTNOSE JACK

Young Bluntnose Jack

20 in.

LESSER AMBERJACK

12 in.

GREATER AMBERJACK

5 ft.

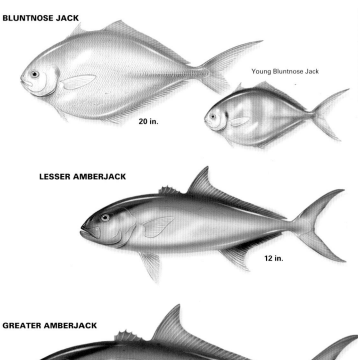
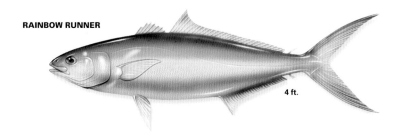
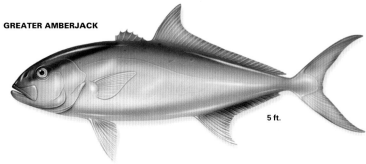

DOLPHINS, LOUVAR, OPAH

Dolphins are colorful pelagic predators that are highly prized for their delicious meat. They are often found near or under floating debris or Gulfweed mats. Louvars and Opahs are unusual solitary pelagic fish.

DOLPHIN (DORADO)
Coryphaena hippurus

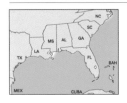

Brilliant coloration of greenish blues with gold flecking. Very steep forehead in male (bull). Swift swimmer, hunting in "packs." Feeds on fish and squid. Attracted to floating debris under which it feeds. Excellent eating. This is the Mahi-Mahi of Hawaii. **Range:** Nova Scotia to Brazil. **Size:** To 7 ft. (2.1 m).

POMPANO DOLPHIN
Coryphaena equiselis

Separated from very similar male Dolphin by smaller size, but may be confused with young or female Dolphin. Silvery to pale yellow. Note small pectoral fin and convex (not concave) edge of anal fin. **Range:** New Jersey to Brazil. **Size:** To 7 ft. (2.1 m).

LOUVAR
Luvarus imperialis

Flattened silver fish with red to pink fins. Steep forehead in adult. Appearance changes with age; young specimens look like puffers. May be related to tunas. Solitary, rarely seen. **Range:** Temperate waters worldwide. Prefers deep water at edge of continental shelf. **Size:** To 6 ft. (1.8 m).

OPAH
Lampris guttatus

Almost flat, circular in form, with stiff, long pectoral fins with sharp tips. Note upward arc of lateral line above pectoral fin. Bright red fins contrast with spotted, bluish silver body. Jaws lack teeth. Important food fish in Europe. **Range:** Grand Banks to Florida, Gulf of Mexico, and Caribbean. **Size:** To 6 ft. (1.8 m).

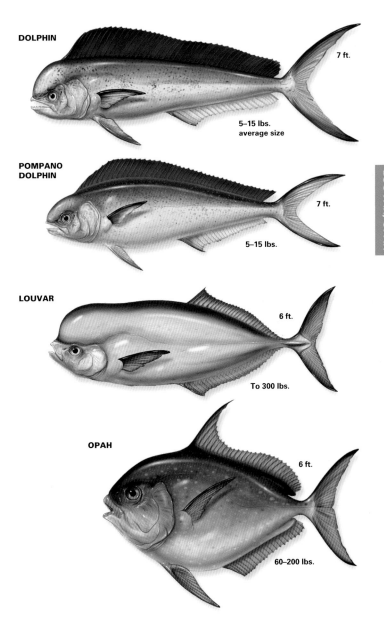

DOLPHIN

7 ft.

5–15 lbs.
average size

**POMPANO
DOLPHIN**

7 ft.

5–15 lbs.

LOUVAR

6 ft.

To 300 lbs.

OPAH

6 ft.

60–200 lbs.

SNAPPERS

QUEEN SNAPPER
Etelis oculatus

Deep red, with elegant, flowing tail fins with tips that continue to elongate with age. Favors deep waters, but will come up to shallow waters over reefs. **Range:** Cape Hatteras to Florida and Gulf of Mexico. **Size:** To 20 in. (50 cm).

VERMILLION SNAPPER
Rhomboplites aurorubens

Overall reddish on upperparts fading to pink below. Very similar to the Red Snapper (see page 122), but note the more rounded anal fin. Favors inshore shoals, ledges, and rocky outcrops. **Range:** North Carolina south to Brazil and Gulf of Mexico. **Size:** To 22 in. (56 cm).

YELLOWTAIL SNAPPER
Ocyurus chrysurus

Broad yellow-green line runs along entire flank from in front of eye to all-yellow tail. Bluish gray upper flank lined with yellow angular bars. Found inshore and offshore over sandy bottoms and reefs. **Range:** Massachusetts to Brazil and Gulf of Mexico. **Size:** To 30 in. (76 cm).

GRAY SNAPPER
Lutjanus griseus

Dark green to grayish above with a rusty tint. Black line runs from eye to dorsal fin. Can change colors from light to dark, especially when in a feeding frenzy. Very similar to the Cubera Snapper. One sure separation is to look at the vomarine teeth in the roof of the mouth. In the Gray Snapper they form an anchor shape; in the Cubera, the shape is pyramidal. Common around mangroves as well as offshore over reefs. **Range:** Massachusetts to Rio de Janerio and Gulf of Mexico. **Size:** To 36 in. (91 cm).

SCHOOLMASTER
Lutjanus apodus

Deep bodied with side bars. In greater size these bars fade and the base color is grayish with yellow fins. Inshore waters especially near elkhorn coral reefs. **Range:** Florida to Brazil and Gulf of Mexico. Occasionally wanders farther north. **Size:** To 24 in. (61 cm).

CUBERA SNAPPER
Lutjanus cyanopterus

Very large—at times confused with groupers. Dark above, grayish below. Shows side blotches. Vomarine teeth are pyramidal in shape, as opposed to the anchor-shaped teeth of large Gray Snappers. Favors inshore and offshore reefs, occasionally wandering into bays and even creeks when young. **Range:** Florida to Brazil and Gulf of Mexico. **Size:** To 5 ft. (1.5 m). **Red List – Vulnerable**

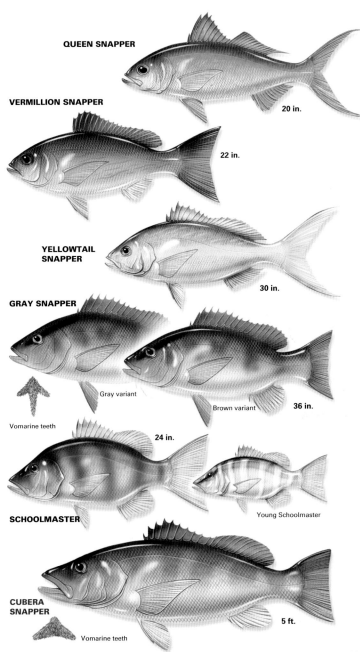

QUEEN SNAPPER

20 in.

VERMILLION SNAPPER

22 in.

YELLOWTAIL
SNAPPER

30 in.

GRAY SNAPPER

Gray variant

Brown variant

36 in.

Vomarine teeth

24 in.

Young Schoolmaster

SCHOOLMASTER

CUBERA
SNAPPER

5 ft.

Vomarine teeth

SNAPPERS

MUTTON SNAPPER
Lutjanus analis

"False eye" spot at dorsal fin base, seen in other snappers as well, perhaps for deflecting predator attacks to the head. Distinct blue stripes on gills. Younger fish have wide side barring. As with most snappers, young favor inshore habitats, and adults are found offshore on reefs or rocky outcrops. **Range:** Massachusetts to Brazil and Gulf of Mexico. **Size:** To 30 in. (76 cm).

SILK SNAPPER
Lutjanus vivanus

Deep rusty red with distinctive yellow eye. Favors deep waters, at depths of 600 ft. (183 m) or more. **Range:** North Carolina to Brazil, most common in Caribbean waters. **Size:** To 30 in. (76 cm).

RED SNAPPER
Lutjanus campechanus

A favorite food fish. Beautiful pinkish red intensifies as this snapper grows in size. Deep red eye. Note triangular shape of anal fin. Found inshore and offshore on rocky bottoms and reefs. When inshore waters cool, Red Snappers move offshore. **Range:** Massachusetts to Brazil and Gulf of Mexico. **Size:** To 31 in. (79 cm).

DOG SNAPPER
Lutjanus jocu

Pinkish to rusty color. Note bluish dotted line below the eye. Large canine teeth can be see when the mouth is closed, hence its name. Found in inshore and offshore reefs and rocky bottoms. **Range:** Rare north of Florida, south to Brazil and Gulf of Mexico. **Size:** To 29 in. (74 cm).

LANE SNAPPER
Lutjanus synagris

Pinkish with distinctive bands of yellow flank markings. Large dark spot below yellowish dorsal fin. Found on offshore reefs, but comes to inshore waters and sea grass beds to feed on shrimp. **Range:** North Carolina to Brazil and Gulf of Mexico. **Size:** To 20 in. (51 cm).

BLACKFIN SNAPPER
Lutjanus buccanella

Pink overall with distinct black crescent spot on base of pectoral fin. Forms large schools at inshore and offshore reefs and ledges. **Range:** North Carolina to Brazil, Caribbean, and Gulf of Mexico. **Size:** To 12 in. (30 cm).

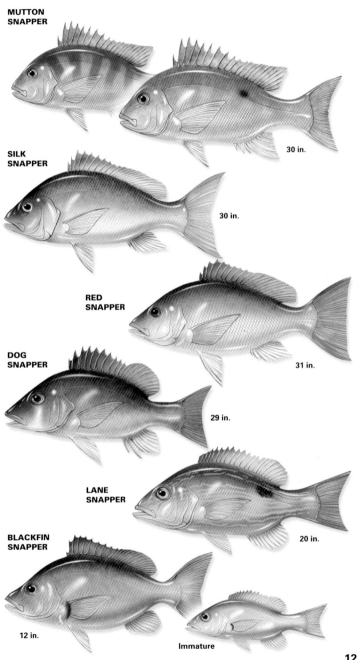

MUTTON SNAPPER

30 in.

SILK SNAPPER

30 in.

RED SNAPPER

31 in.

DOG SNAPPER

29 in.

LANE SNAPPER

20 in.

BLACKFIN SNAPPER

12 in.

Immature

GRUNTS

BLACK MARGATE
Anisotremus surinamensis

Silvery gray with charcoal blotching on side and across forehead. Favors jetties and rocky outcrops of inshore shallows more than offshore sites over reefs. **Range:** South Carolina to Bermuda south to Florida and Gulf of Mexico. **Size:** To 14 in. (36 cm).

PORKFISH
Anisotremus virginicus

Boldly patterned: two black stripes on foreparts, yellow to orange stripes along body, and yellow to orange fins. Found in shallow waters over reefs, where it can form large schools. **Range:** Florida to West Indies and eastern portion of southern Gulf of Mexico. **Size:** To 14 in. (36cm).

BLUESTRIPED GRUNT
Haemulon sciurus

Streamlined, all-yellow body lined with blue stripes. Rear portion of dorsal fin and base of anal fin may be dark. Favors shallow inshore reefs, where it forms large schools. May move into coastal channels. **Range:** Rare north to North Carolina, most common from southern Florida to Brazil. **Size:** To 10 in. (25 cm).

FRENCH GRUNT
Haemulon flavolineatum

Small, yellowish body covered with blue stripes. Fins all yellow. Large, sharply pointed pectoral fin. Found in shallow-water reefs and inshore channels. Can be abundant off south Florida and Florida Keys, particularly on coral reefs. **Range:** North Carolina (uncommon) south to Brazil. **Size:** To 8 in. (20 cm).

MARGATE
Haemulon album

Plainly colored in comparison to other grunts. Deep bodied, bluish gray with darker fins. No outstanding markings. Found from offshore reefs to inshore shallow-water coral bottoms. **Range:** Florida and West Indies. **Size:** To 24 in. (61 cm).

TOMTATE
Haemulon aurolineatum

Silvery blue to gray with yellow stripes that vary in thickness. Usually shows dark spot on peduncle, but variants lack tail spot. Ranges from sandy bottoms and grass beds to offshore reefs. **Range:** North Carolina to Brazil and Gulf of Mexico, more common toward south. **Size:** To 10 in. (25 cm).

WHITE GRUNT
Haemulon plumieri

Large, with variable color. Most are silver-gray with a concentration of yellow and blue lines on head and gills. Found offshore and inshore over areas of rocky and coral bottoms, as well as on coral reefs and inshore marine grass beds on sandy flats. **Range:** South Florida and Gulf of Mexico south to Brazil. **Size:** To 24 in. (61 cm).

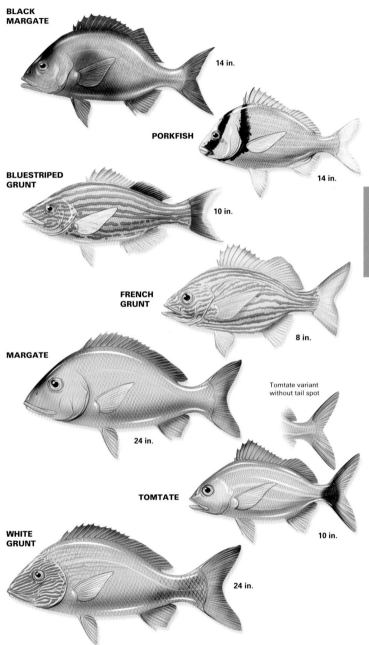

BLACK MARGATE

14 in.

PORKFISH

14 in.

BLUESTRIPED GRUNT

10 in.

FRENCH GRUNT

8 in.

MARGATE

24 in.

Tomtate variant without tail spot

TOMTATE

10 in.

WHITE GRUNT

24 in.

PORGIES

JOLTHEAD PORGY
Calamus bajonado

One of the largest porgies. Note sharply sloped forehead and eye surrounded by blue. Dark smudge speckles on flanks. Feeds on sea urchins and crabs in inshore waters over coral and sand bottoms. **Range:** North to Rhode Island, though less common in northern waters, and south to Brazil and Gulf of Mexico. **Size:** To 27 in. (69 cm).

RED PORGY
Pagrus pagrus

Reddish in color with flat forehead. Tiny blue flecks on sides, blue surrounds eye. Deepwater species found well offshore. **Range:** New York to Argentina and Gulf of Mexico. **Size:** To 16 in. (41cm). **Red List – Endangered**

SPOTTAIL PINFISH
Diplodus holbrooki

Lacks flat forehead typical of other porgies. Rounded in shape with side banding and black tail peduncle. Dorsal fin spines are prominent. Found in inshore shallows to offshore. **Range:** North Carolina south into Gulf of Mexico. **Size:** To 8 in. (20 cm).

KNOBBED PORGY
Calamus nodosus

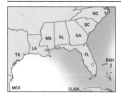

Sharply sloped forehead. Blue face and reddish cheeks make this porgy quite colorful. Found over coral and sand bottoms. **Range:** North Carolina to Florida and eastern Gulf of Mexico. **Size:** To 16 in. (41 cm).

PINFISH
Lagodon rhomboides

Named for sharp spines on dorsal fin. Greenish above fading to silver below. Flanks with yellow lines and dark bars. Distinct dark spot behind gills. Favors shallow waters with aquatic vegetation. **Range:** Massachusetts south to Florida and Gulf of Mexico. **Size:** To 8 in. (20 cm).

SHEEPSHEAD PORGY
Calamus penna

Blunt head with large teeth protruding from mouth. Distinct dark side banding from head to tail. Dark face. An inshore fish of bays and shallows and inshore reefs. Prefers muddy bottoms over oyster beds where it forages, and also around pilings and bridge abutments. **Range:** Massachusetts to Brazil and Gulf of Mexico, absent from Caribbean. **Size:** To 30 in. (76 cm).

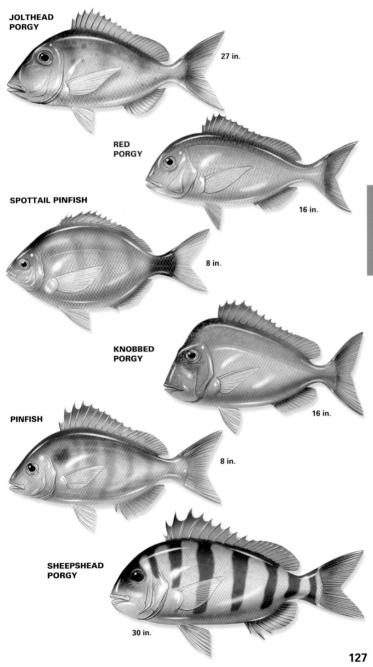

JOLTHEAD PORGY — 27 in.

RED PORGY — 16 in.

SPOTTAIL PINFISH — 8 in.

KNOBBED PORGY — 16 in.

PINFISH — 8 in.

SHEEPSHEAD PORGY — 30 in.

127

PORGIES

WHITEBONE PORGY
Calamus leucosteus

Silvery gray with scattered rusty spots covering flanks. Prefers sandy bottoms both inshore and offshore, as well as rocky and reef areas. **Range:** North Carolina to Florida and into the Gulf of Mexico. **Size:** To 18 in. (46 cm).

SILVER PORGY
Diplodus argenteus

Silvery gray with faint, dusky side bars. Distinct dark spot on peduncle. Found in surf areas, sea grass beds, and with rocks and reefs in shallow waters. **Range:** South Florida to Argentina. **Size:** To 8 in. (20 cm).

SEA BREAM
Archosargus rhomboidalis

Silvery with yellow stripes running length of body. Dark spot behind gill cover. Prefers shallow inshore waters near jetties, rocky areas, and hard-bottom areas. **Range:** New Jersey to Argentina, including northeastern Gulf of Mexico but not Bahamas. **Size:** To 14 in. (36 cm).

GRASS PORGY
Calamus arctifrons

Silvery with vertical dark bars and dark bar through eye: excellent camouflage in sea grass areas. Found on grassy bottoms of inshore shallows. **Range:** South Florida and southeastern Gulf of Mexico. **Size:** To 10 in. (25 cm).

LITTLEHEAD PORGY
Calamus proridens

Silvery with gold flecking. Prefers inshore waters with sandy bottoms and reef areas. **Range:** North Carolina south to Florida and into Gulf of Mexico. **Size:** To 18 in. (46 cm).

SAUCEREYE PORGY
Calamus calamus

Yellowish with bluish fins and distinct blue lines near eye (making it look large) and on gill cover. Blue-tinted fins. Found from sandy and coral bottoms with vegetation inshore to offshore waters and reefs. **Range:** North Carolina south through Florida, into Gulf of Mexico and West Indies. **Size:** To 16 in. (41 cm).

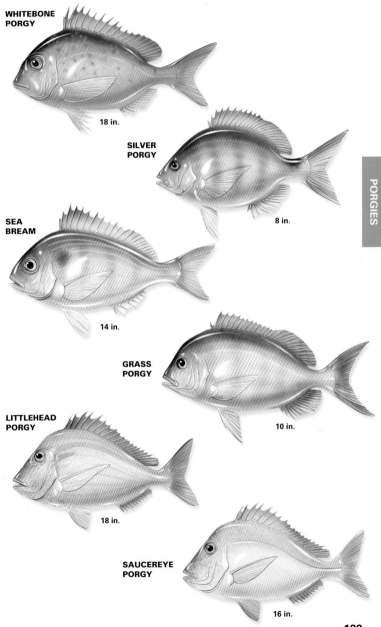

WHITEBONE PORGY

18 in.

SILVER PORGY

8 in.

SEA BREAM

14 in.

GRASS PORGY

10 in.

LITTLEHEAD PORGY

18 in.

SAUCEREYE PORGY

16 in.

PORGIES

SCUP (NORTHERN PORGY) *Stenotomus chrysops*

Gray-silver, with nearly oval body profile and irregularly shaped, faint bars on flanks. Bars are sometimes absent. Ranges through almost all coastal environments, from protected bays and inlets to offshore reefs and deeper water. **Range:** Nova Scotia to Cape Hatteras and stragglers father south. **Size:** To 18 in. (46 cm).

BLACK SEA BASS *Centropristis striata*

Dark gray or gray-brown, often (but not always) with indistinct bars or darker patches along flanks. Pale-centered scales give flanks an extra "scaly" appearance. Ranges widely though all coastal environments and into deeper waters. **Range:** Maine to northern Florida and northern Gulf of Mexico, but uncommon south of South Carolina and Tampa. **Size:** To 24 in. (61 cm).

CUNNER *Tautogolabrus adspersus*

Color varies widely from red-brown to dark greenish gray. May also have darker bars or irregular patches on flanks. Snout longer than the closely related Tautog. Found inshore around docks, pilings, and other structures and in deeper water over rocky bottoms. **Range:** Newfoundland south of Cape Hatteras, with stragglers a little farther south. **Size:** To 15 in. (38 cm).

TAUTOG *Tautoga onitis*

Stocky, dull gray-green to gray-brown, with blunt snout and flanks covered with dark, irregular blotches and spots. Adult males are typically the darkest and have more protruding lips. Favors docks, pilings, breakwaters, and open waters over rocky bottoms. **Range:** Nova Scotia to South Carolina. **Size:** To 15 in. (38 cm).

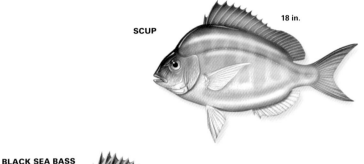

SCUP

18 in.

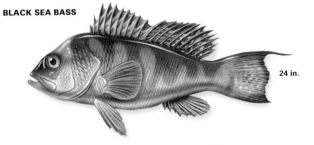

BLACK SEA BASS

24 in.

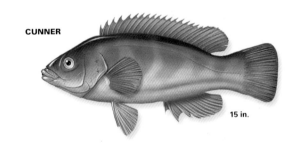

CUNNER

15 in.

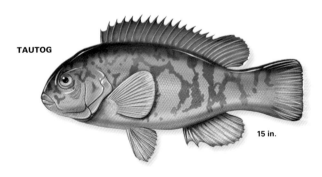

TAUTOG

15 in.

DRUMS

BLACK DRUM   *Pogonias cromis*

Stout, dark body, with dark fins and diffuse dark bands along flanks. Many chin barbels. Usually found over muddy or sandy bottoms close inshore or in surf along beaches. **Range:** Nova Scotia to Florida and Gulf of Mexico. **Size:** To 5.5 ft. (1.7 m).

RED DRUM   *Sciaenops ocellatus*

Red-brown or bronze, with an elongated body. Virtually all Red Drums have one or more spots just ahead of the tail. Young favor grassy shallows or shell beds close inshore, but adults move to deeper offshore waters. **Range:** Massachusetts to Florida and Gulf of Mexico. **Size:** To 5 ft. (1.5 m).

SPOT   *Leiostomus xanthurus*

Compact, oval body with brassy red upper flanks shading to silver-gray below. Dark diagonal stripes across flanks. Only drum with a forked tail. Note the namesake spot just behind the gill cover. Common in surf along beaches, and in bays and estuaries. **Range:** Massachusetts to Florida and northern Gulf of Mexico, but uncommon in southern Florida and Florida Keys. **Size:** To 15 in. (38 cm).

ATLANTIC CROAKER   *Micropogonias undulatus*

A silver drum with diagonal black stripes running across flanks. Several small barbels normally present under chin. Prefers sandy or shell areas in surf, as well as more protected shallow bays and estuaries. **Range:** Massachusetts to Florida, and throughout Gulf of Mexico, but uncommon in southern Florida and Florida Keys. **Size:** To 21 in. (53 cm).

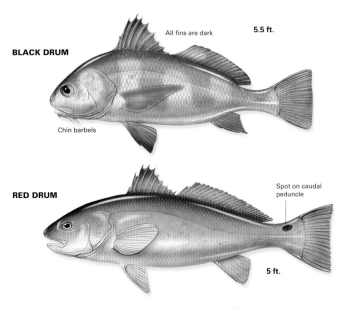

BLACK DRUM

All fins are dark

5.5 ft.

Chin barbels

RED DRUM

Spot on caudal peduncle

5 ft.

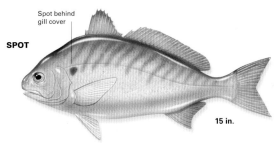

Spot behind gill cover

SPOT

15 in.

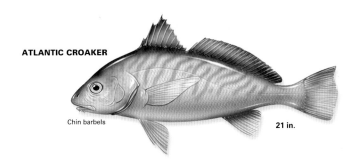

ATLANTIC CROAKER

Chin barbels

21 in.

UNUSUAL DRUMS, GOATFISH

HIGH-HAT *Pareques acuminatus*

Unmistakable tropical drum. Bold zebra stripes, unusual body profile. Shy inhabitant of coral reefs and rocky areas, where it prefers sheltered nooks that offer a quick retreat to safety. **Range:** Florida, Bahamas, and Caribbean. **Size:** To 9 in. (23 cm).

JACKKNIFE FISH *Equetus lanceolatus*

Boldly marked yellow drum with black stripes and elongated dorsal fin. Shy inhabitant of coral reefs and other areas that offer ready shelter. This less common tropical drum feeds and roams primarily at night. **Range:** North Carolina to Florida, Gulf of Mexico, and Caribbean. **Size:** To 11 in. (28 cm).

SPOTTED DRUM *Equetus punctatus*

Another boldly marked, unmistakable tropical drum. Same habits as the other reef drums here: a shy fish that favors sheltered nooks in the reef and ranges out mostly at night. **Range:** Florida and most of Gulf Coast and Caribbean, but more common in warmer southern waters. **Size:** To 10 in. (25 cm).

YELLOW GOATFISH *Mulloidichthys martinicus*

A pale silvery fish with a yellow stripe along flanks from eye to tail fin. Long barbels under chin. A very common inhabitant of coral reefs, where dense schools of goatfish often shelter under overhangs of coral. **Range:** Florida and the Bahamas to the Gulf of Mexico and Caribbean. **Size:** To 12 in. (30 cm).

SPOTTED GOATFISH *Pseudupeneus maculatus*

Common silver-gray goatfish with two or three dark spots along flanks. Two prominent barbels project down from chin. **Range:** From New Jersey south, but unusual north of Florida. Abundant on Florida, Gulf of Mexico, and Caribbean reefs. **Size:** To 11 in. (28 cm).

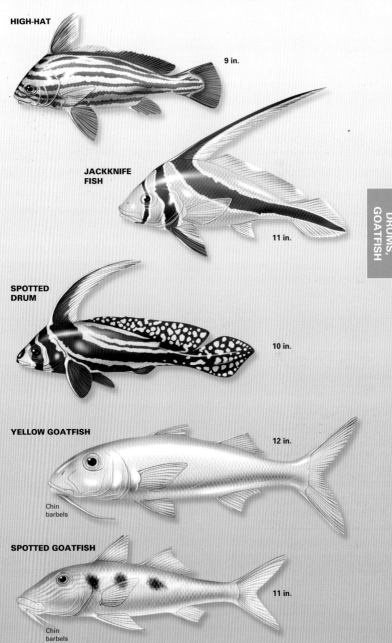

HIGH-HAT

9 in.

JACKKNIFE FISH

11 in.

SPOTTED DRUM

10 in.

YELLOW GOATFISH

12 in.

Chin barbels

SPOTTED GOATFISH

11 in.

Chin barbels

KINGFISH, MULLETS

NORTHERN KINGFISH
Menticirrhus saxatilis

Gray-brown to brown with irregular splotches on flanks and fins. First ray of dorsal fin projects prominently. Short barbels under chin. **Range:** Massachusetts to Florida and Gulf of Mexico, but less common south of Cape Hatteras. **Size:** To 18 in. (46 cm).

SOUTHERN KINGFISH
Menticirrhus americanus

Similar to the Northern Kingfish, but lighter in color, with silvery gray flanks and long diagonal stripes. No long extension on first ray of dorsal fin. Short chin barbels. **Range:** Long Island to Florida and Gulf of Mexico, but unusual in southern Florida and Bahamas. Most abundant along mid-Atlantic Coast. **Size:** To 15 in. (38 cm).

GULF KINGFISH
Menticirrhus littoralis

Similar in profile to the Southern Kingfish, but with blue to silvery white flanks that do not have markings or stripes. Tail fin edged with dark band. Found in surf along beaches and over sandy bottoms and shallow flats. **Range:** Virginia to Florida and Gulf Coast, absent from southern Florida. **Size:** To 18 in. (46 cm).

WHITE MULLET
Mugil curema

Sleek, silvery, with blue-green back and unmarked flanks, except for dark blotch at base of pectoral fin. Roams widely in shallow coastal waters and enters freshwater rivers. The most abundant mullet of warm southern waters. **Range:** Massachusetts to Florida and Gulf of Mexico. **Size:** To 35 in. (89 cm), but usually much smaller.

STRIPED MULLET
Mugil cephalus

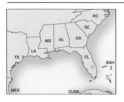

Elongated, silvery, with dark blue or dark green back and unmarked flanks. The most commercially important mullet species, widely fished for food along the southeastern and Gulf coasts. **Range:** Nova Scotia to Florida and the Gulf of Mexico, most abundant south of Cape Hatteras and in Gulf. **Size:** To 36 in. (91 cm), but usually much smaller.

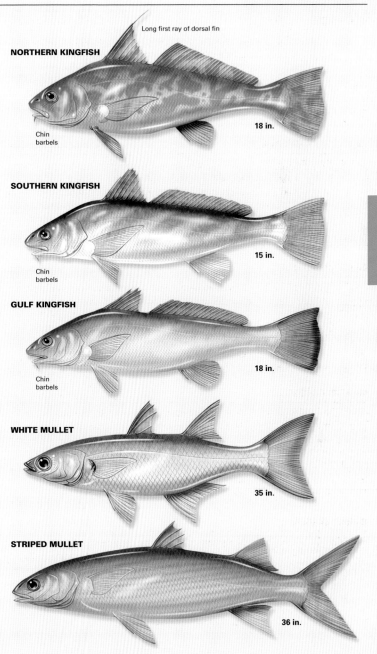

Long first ray of dorsal fin

NORTHERN KINGFISH

Chin barbels

18 in.

SOUTHERN KINGFISH

Chin barbels

15 in.

GULF KINGFISH

Chin barbels

18 in.

WHITE MULLET

35 in.

STRIPED MULLET

36 in.

WEAKFISH, SEA TROUT, PERCH

WEAKFISH
Cynoscion regalis

Weak mouth parts allow this popular sport fish to toss hooks easily, hence its name. Prefers deeper coastal bays and channels but wanders into shallows when hunting. Also occurs in surf along beaches. **Range:** Nova Scotia to northern Florida. **Size:** To 36 in. (91 cm).

SPOTTED SEATROUT
Cynoscion nebulosus

Bluish to greenish gray above grading to silver-white on belly. Flanks and upper fins covered with pattern of irregular dark spots. A popular sport and food fish that prefers grassy flats and shallows when hunting but moves offshore in cold weather. **Range:** Long Island to Florida and Gulf Coast. **Size:** To 36 in. (91 cm).

SAND SEATROUT
Cynoscion arenarius

Similar in overall form to the Spotted Seatrout, but smaller and without noticeable spots or patches. Typically pale silver-white with darker brown or bronze-tinted back. Prefers deeper water but wanders into shallow flats to feed. **Range:** Florida and Gulf Coast. **Size:** To 16 in. (41 cm).

SILVER SEATROUT
Cynoscion nothus

Difficult to distinguish from the larger Sand Seatrout, except by range north of Florida. Typically more silver-bluish, with fine diagonal rows of spots or subtle stripes across upper flanks. **Range:** Chesapeake Bay to Cape Canaveral, and Tampa through Gulf Coast. **Size:** To 12 in. (30 cm).

WHITE PERCH
Morone americanus

A common species that often migrates up rivers with other fish, such as Alewife, to breed. A popular recreational and important commercial fish. **Range:** Nova Scotia to North Carolina. **Size:** To 19 in. (48 cm).

WEAKFISH

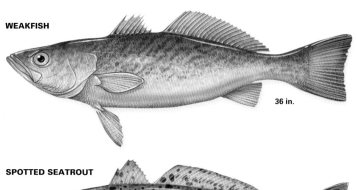

36 in.

SPOTTED SEATROUT

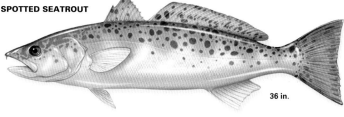

36 in.

SAND SEATROUT

16 in.

SILVER SEATROUT

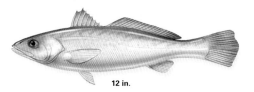

12 in.

WHITE PERCH

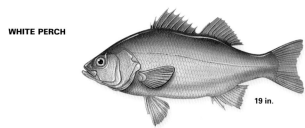

19 in.

139

LINED SEAHORSE
Hippocampus erectus

The most common seahorse, but often overlooked because it is slow-moving and very well camouflaged. Color and pattern is highly variable. Found in mats of Gulfweed (*Sargassum*) and other cover, such as marine grass beds. **Range:** Nova Scotia to Argentina, Gulf of Mexico. **Size:** To 6 in. (15 cm).

LONGSNOUT SEAHORSE
Hippocampus reidi

Body usually brownish yellow, scattered with many distinct dark brown spots. Snout longer and thinner than that of the Lined Seahorse. Uncommon throughout its range. **Range:** Cape Hatteras to Brazil, Bermuda. **Size:** To 6 in. (15 cm).

DWARF SEAHORSE
Hippocampus zosterae

Tiny, with plain flanks without spots or other markings. Color ranges from light brown to dark browns and greens. Common in seagrass beds, but shy and inconspicuous, so it is easy to overlook. **Range:** South Florida, Bahamas, and Gulf of Mexico. **Size:** To 2 in. (5 cm).

SARGASSUMFISH
Histrio histrio

A small fish usually found embedded in mats of Gulfweed (*Sargassum*) algae well offshore. Thoroughly camouflaged, with bold markings that combine to make it almost invisible within the Gulfweed. **Range:** Open ocean, but sometimes blown inshore within floating mats of algae. **Size:** To 8 in. (20 cm).

LONGLURE FROGFISH
Antennarius multiocellatus

An unusual bottom-dwelling fish adapted to use sponge species as camouflage. Color and pattern highly variable to match the local sponges. Found in coral reefs and anywhere sponges are common. Long nose extension acts as lure for its prey of small fish. **Range:** Florida, Bahamas, and Caribbean. **Size:** To 8 in. (20 cm).

SPOTTED SCORPIONFISH
Scorpaena plumieri

An elongated bottom-dwelling fish with bold, complex camouflage patterns, usually in reds and red-brown with lighter spots and blotches. Inconspicuous on sand floor of coral reef areas. *First few spines of dorsal fin are venomous.* **Range:** New York to Florida, northern Gulf of Mexico. **Size:** To 12 in. (30 cm).

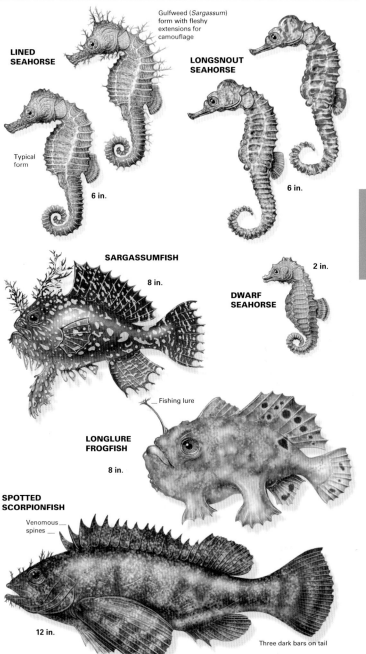

LINED SEAHORSE

Gulfweed (*Sargassum*) form with fleshy extensions for camouflage

Typical form

LONGSNOUT SEAHORSE

6 in.

6 in.

SARGASSUMFISH

8 in.

2 in.

DWARF SEAHORSE

Fishing lure

LONGLURE FROGFISH

8 in.

SPOTTED SCORPIONFISH

Venomous spines

12 in.

Three dark bars on tail

141

TOADFISH, CATFISH, GOOSEFISH

OYSTER TOADFISH
Opsanus tau

Toadfish are squat, highly camouflaged bottom-dwellers and hunt their prey by ambush. Dark brown and gray, with heavy, complex pattern of markings. Favors rocky bottoms and reefs in shallow water. **Range:** Cape Cod to northern Florida. **Size:** To 14 in. (36 cm).

GULF TOADFISH
Opsanus beta

Similar to the Oyster Toadfish in habits and ambush hunting methods. Heavily marbled camouflage patterns. Common but inconspicuous resident of shallow grass flats and rocky areas in coastal bays and inlets. **Range:** Cape Canaveral south and Gulf Coast. **Size:** To 12 in. (30 cm).

GAFFTOPSAIL CATFISH
Bagre marinus

Sleek, blue-gray, with sail-like dorsal fin. Long pair of "whisker" barbels extends from corners of mouth, short barbels under chin. Frequents a variety of shallow coastal waters and river mouths. **Range:** Massachusetts to Florida and Gulf Coast. **Size:** To 24 in. (60 cm).

HARDHEAD CATFISH
Arius felis

Brownish gray to blue-gray above grading to light yellow below. Mouth barbels shorter than in the Gafftopsail Catfish, chin barbels longer. Dorsal fin small, without extension. **Range:** Massachusetts to Florida and Gulf Coast. **Size:** To 24 in. (60 cm).

GOOSEFISH
Lophius americanus

An amazingly ugly fish. A relative of the deep-sea anglerfishes, as shown by its highly modified dorsal fin, which arises above the mouth area and acts as a fishing lure for small prey. Typically found in shallow waters in winter, migrating to deeper waters in summer. Sold as "monkfish" in markets. **Range:** Nova Scotia to well south of Cape Hatteras, but only in very deep waters in the south. **Size:** To 4 ft. (1.2 m).

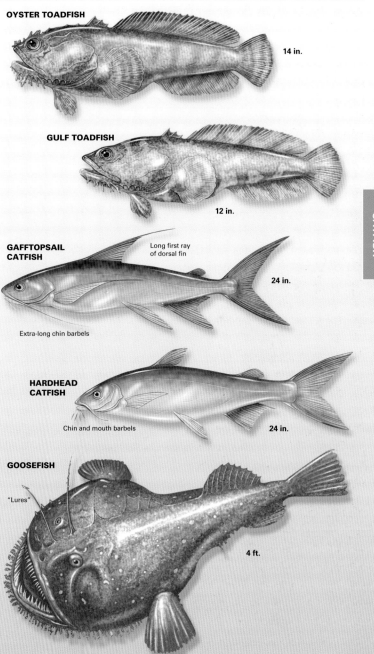

OYSTER TOADFISH

14 in.

GULF TOADFISH

12 in.

GAFFTOPSAIL CATFISH

Long first ray
of dorsal fin

24 in.

Extra-long chin barbels

HARDHEAD CATFISH

Chin and mouth barbels

24 in.

GOOSEFISH

"Lures"

4 ft.

SMALL REEF FISH

BLUE CHROMIS *Chromis cyaneus*

Brilliant blue with dark back and deeply forked tail. Tail fin edged black on upper and lower edges. Very common on coral reefs. **Range:** Florida, Bahamas, Gulf of Mexico, and Caribbean. **Size:** To 5 in. (12.5 cm).

BROWN CHROMIS *Chromis multilineata*

Light green, brown, or gray, tail and edge of dorsal fin tinged with yellow. Dark spot at base of pectoral fin. Common on coral reefs. **Range:** Florida, Bahamas, Gulf of Mexico, and Caribbean. **Size:** To 6.25 in. (16 cm).

YELLOWTAIL DAMSELFISH *Microspathodon chrysurus*

Dark brown overall, sometimes dark blue. Tail brilliant yellow. Intense but tiny blue spots across flanks, particularly in juveniles. Very common on coral reefs. **Range:** Florida, Bahamas, Gulf of Mexico, and Caribbean. **Size:** To 8 in. (20 cm).

SERGEANT MAJOR *Abudefduf saxatilis*

Five dark bars across gray body with yellow back. Very common swimming over and within coral reefs. **Range:** Cape Cod to Florida, Bahamas, Gulf of Mexico, and Caribbean. **Size:** To 7 in. (15 cm).

BICOLOR DAMSELFISH *Pomacentrus partitus*

Body color split sharply between dark front and white to yellow rear. Mix of bright yellow and white areas varies widely. **Range:** Florida, Bahamas, Gulf of Mexico, and Caribbean. **Size:** To 4 in. (10 cm).

COCOA DAMSELFISH *Pomacentrus variabilis*

Brown above, pale to bright yellow below. Tail and pectoral fins mostly yellow. Common on coral reefs. **Range:** Florida, Bahamas, Gulf of Mexico, and Caribbean. **Size:** To 5 in. (12.5 cm).

CARDINAL SOLDIERFISH *Plectrypops retrospinis*

Brilliant red-orange, with spiny dorsal fin. Rounded lobes of fins and tail. **Range:** Florida, Bahamas, Gulf of Mexico, and Caribbean. **Size:** 5 in. (13 cm).

BIGEYE *Priacanthus arenatus*

Bright red-orange to salmon, sometimes pink, sometimes with darker spots or blotches. Large eye. Common on rock and coral reefs. **Range:** Cape Cod to Florida, Bahamas, Gulf of Mexico, and Caribbean. **Size:** To 12 in. (30 cm).

BLACKBAR SOLDIERFISH *Myripristis jacobus*

Red to pale red grading to pink below. Dark vertical bar just behind gill cover. Common in large schools on coral reefs. **Range:** North Carolina to Florida, Bahamas, Gulf of Mexico, and Caribbean. **Size:** To 8 in. (20 cm).

SQUIRRELFISH *Holocentrus adscensionis*

Dull red, sometimes bright red, or pink grading to silver white. Body marked with diffuse stripes or blotches. Fins are pale pink. Common. **Range:** Florida, Bahamas, Gulf of Mexico, and Caribbean. **Size:** To 12 in. (30 cm).

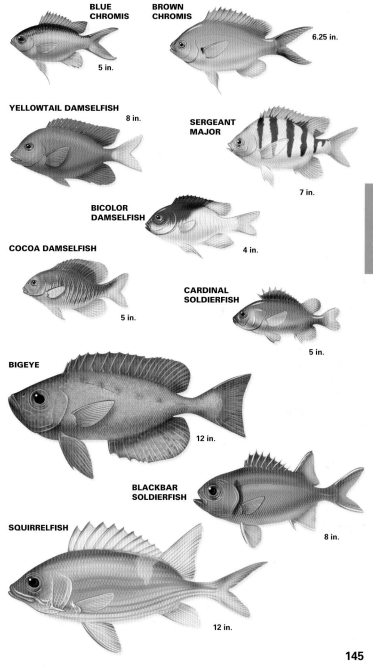

BLUE CHROMIS

BROWN CHROMIS

6.25 in.

5 in.

YELLOWTAIL DAMSELFISH

8 in.

SERGEANT MAJOR

7 in.

BICOLOR DAMSELFISH

4 in.

COCOA DAMSELFISH

CARDINAL SOLDIERFISH

5 in.

5 in.

BIGEYE

12 in.

BLACKBAR SOLDIERFISH

SQUIRRELFISH

8 in.

12 in.

145

BUTTERFLYFISH

Chaetodon sp.

Basically a tropical group, but most species occasionally stray north to Virginia or even Long Island or Cape Cod waters.

BANK BUTTERFLYFISH

Chaetodon aya

Tan body marked by two distinct black bands, one from leading edge of dorsal fin to eye. Long snout. No heavy bar below eye. Coastal species that prefers shallow waters of underwater ledges and coral reefs. **Range:** North Carolina to Florida, Gulf of Mexico, and Caribbean. **Size:** To 6 in. (15 cm).

BANDED BUTTERFLYFISH

Chaetodon striatus

Blue-gray with four distinct bands. Snout not elongate. Caudal, dorsal, and anal fins edged with teal blue. Juvenile has same basic pattern but with large black spot at rear of dorsal fin. **Range:** Southern New Jersey to Florida, Gulf of Mexico, and Caribbean. **Size:** To 6 in. (15 cm).

REEF BUTTERFLYFISH

Chaetodon sedentarius

Two dark bands, one running across eye. Flanks yellow above fading to silver-white below. Common on coral reefs and in shallow coastal waters. **Range:** South Florida and Bahamas through Caribbean. **Size:** To 6 in. (15 cm).

SPOTFIN BUTTERFLYFISH

Chaetodon ocellatus

Large, single-banded. Base color is white tinted with yellow. Fairly long snout. Fins all yellow with markings. Rear of extensive dorsal fin marked with black spot. Found on coral reefs to inshore waters. **Range:** Bahamas, Florida coastal waters, and Caribbean. **Size:** To 8 in. (20 cm).

LONGSNOUT BUTTERFLYFISH

Chaetodon aculeatus

Small with very long snout. Flanks bright yellow above fading to silver-white below. Fairly common, but shy of divers and likely to hide on approach. **Range:** South Florida, Bahamas, Gulf of Mexico, and Caribbean. **Size:** To 2–3 in. (5–8 cm).

FOUREYE BUTTERFLYFISH

Chaetodon capistratus

Large black "eye" spot just forward of the tail, and a pale eye stripe. Body silver-white fading to yellow below and to blue on rear dorsal fin and tail. Common on coral reefs, inshore tropical waters. **Range:** Florida, Bahamas, Gulf of Mexico, and Caribbean. **Size:** To 6 in. (15 cm).

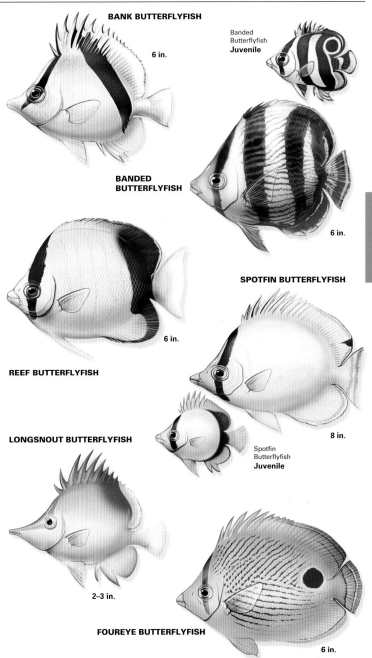

BANK BUTTERFLYFISH

6 in.

Banded
Butterflyfish
Juvenile

**BANDED
BUTTERFLYFISH**

6 in.

SPOTFIN BUTTERFLYFISH

REEF BUTTERFLYFISH

6 in.

8 in.

Spotfin
Butterflyfish
Juvenile

LONGSNOUT BUTTERFLYFISH

2–3 in.

FOUREYE BUTTERFLYFISH

6 in.

BUTTERFLYFISH

147

SURGEONFISH, CHUBS

OCEAN SURGEON
Acanthurus bahianus

Nearly oval body profile with uniform light blue, gray-blue, or pale brown color on flanks. Fins blue, marked with fine bands of lighter and dark colors near edges. Note sharp spines on either side of tail base. Common over rock and coral reefs and other bottoms that offer shelter. **Range:** Cape Cod to Florida, Bahamas, and Gulf of Mexico except northeast corner. **Size:** To 12 in. (30 cm).

BLUE TANG
Acanthurus coeruleus

Uniform powder blue color ranging to deep blue or violet. A very fine linear pattern can be seen in close views. Eye color varies from bright blue to yellow. Sharp spines at base of tail. Juveniles are bright yellow. Very common on rock and coral reefs. **Range:** Long Island to Florida, Bahamas, Gulf of Mexico, and Caribbean. **Size:** To 10 in. (25 cm).

DOCTORFISH
Acanthurus chirurgus

Very similar to the Ocean Surgeon, but with dark vertical bars along flanks. Dorsal and anal fins are often more intensely colored than in the Ocean Surgeon. Spines at base of tail. Very common on rock and coral reefs. **Range:** Cape Cod to Florida, Bahamas, and Gulf of Mexico. **Size:** To 10 in. (25 cm).

BERMUDA CHUB
Kyphosus sectatrix

YELLOW CHUB
Kyphosus incisor

These two chubs are so similar that separating them visually in the field is impossible. Both have oval, "football-shaped," gray-blue bodies with a fine pattern of yellow horizontal stripes. Faces are gray-blue with yellow and white patches just below the eye. Yellow stripes on the Bermuda Chub are slightly paler than those on the Yellow Chub. Both species are very common, typically seen in schools around reefs. **Range:** Cape Cod to Florida, Bahamas, Gulf of Mexico, and Caribbean. **Size:** 12–14 in. (30–36 cm).

SURGEONFISH, CHUBS

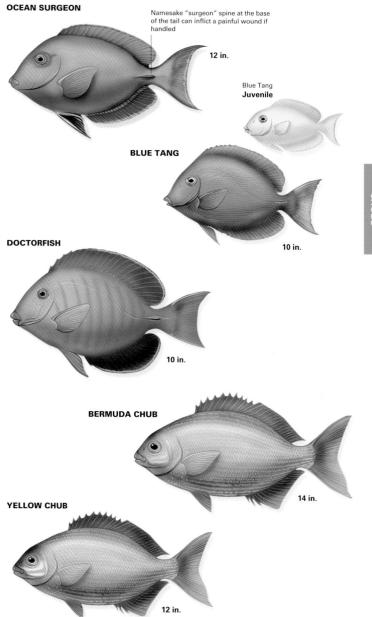

OCEAN SURGEON

Namesake "surgeon" spine at the base of the tail can inflict a painful wound if handled

12 in.

Blue Tang
Juvenile

BLUE TANG

10 in.

DOCTORFISH

10 in.

BERMUDA CHUB

14 in.

YELLOW CHUB

12 in.

SURGEONFISH, CHUBS

149

QUEEN ANGELFISH
Holacanthus ciliaris

Medium-sized, greenish blue body with deep blue outlining dorsal and anal fins. Juvenile very similar to the juvenile Blue Angelfish, both being bluish, striped, and yellow-tailed. **Range:** Bahamas, Florida, Gulf of Mexico, and Caribbean. **Size:** To 12 in. (30cm).

BLUE ANGELFISH
Holocanthus bermudensis

Body scales bluish cream with yellow edges, face bluish. Caudal, dorsal, and anal fins have yellow margins. Unmarked gill cover. Last vertical line on flank of juvenile is straight, not curved as in the juvenile Queen Angelfish. Found in coral reefs. **Range:** North Carolina to Florida, Gulf of Mexico, and Caribbean. **Size:** To 18 in. (45 cm).

GRAY ANGELFISH
Pomacanthus arcuatus

Gray, with thin aqua blue edge to dorsal, anal, and caudal fins. Pale gray mouth region. Juveniles black with vertical yellow bars and yellow caudal fin with black crescent spot. The look-alike juvenile of the French Angelfish has a darker tail band. **Range:** Bahamas, Florida, Gulf of Mexico, and Caribbean. **Size:** To 14 in. (36 cm).

FRENCH ANGELFISH
Pomacanthus paru

Black side scales edged in yellow. Yellow spot at base of pectoral fin, yellow ring around eye. Dorsal and anal fins with ragged edge and long back-facing extensions. Found in shallow waters and coral reefs. Juvenile almost identical to the Gray Angelfish, but note larger black band on tail. **Range:** Roams north to New England, but regular from Carolinas and Bermuda south to Caribbean. **Size:** To 13 in. (33 cm).

ROCK BEAUTY
Holocanthus tricolor

Unmistakable color pattern of black and yellow. Juveniles start out all yellow with small black spot on flank. The black spot spreads with age into adult pattern. Food consists mainly of sponges, some coral, and algae. An inhabitant of reefs and shallow coastal waters. **Range:** Georgia to Florida, Gulf of Mexico, and Caribbean. **Size:** To 12 in. (30 cm).

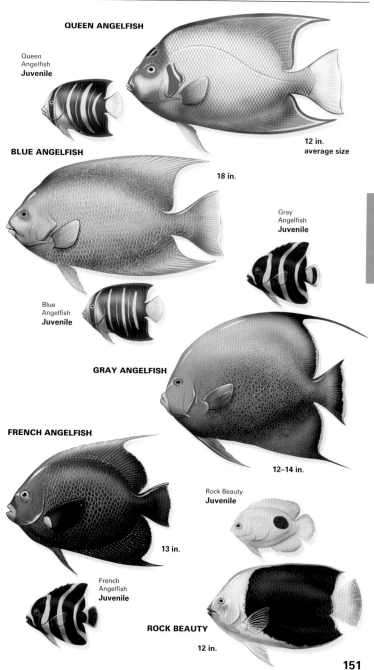

QUEEN ANGELFISH

Queen
Angelfish
Juvenile

12 in.
average size

BLUE ANGELFISH

18 in.

Gray
Angelfish
Juvenile

Blue
Angelfish
Juvenile

GRAY ANGELFISH

12–14 in.

FRENCH ANGELFISH

Rock Beauty
Juvenile

13 in.

French
Angelfish
Juvenile

ROCK BEAUTY

12 in.

SPADEFISH, HOGFISH, TILEFISH

ATLANTIC SPADEFISH
Chaetodipterus faber

Silver-gray to medium gray flanks, with broad vertical stripes. Dorsal and anal fins with long extensions. Occurs in small to large schools around reefs and structures such as piers and offshore platforms. **Range:** Cape Cod to Florida, Bahamas, Gulf of Mexico, and Caribbean. **Size:** To 36 in. (91 cm).

HOGFISH
Lachnolaimus maximus

As in a number of reef species, the Hogfish has several distinct color phases related to sex and age. The intermediate color phase Hogfish is a uniform light red-brown. Juveniles have same basic coloration, often mottled with white patches. In mature, terminal color phase, Hogfish develop a bicolored pattern in which the leading edge of head and back become edged with dark brown or black and the fins and tail develop strong banded color patterns. Mature Hogfish have a long snout, giving the head a distinct sloping profile. **Range:** Nova Scotia to Florida, Bahamas, and Gulf of Mexico. **Size:** To 36 in. (91 cm).

TILEFISH
Lopholatilus chamaeleonticeps

A fleshy protuberance just forward of the dorsal fin is diagnostic of the Tilefish. Elongated body, with blue-gray back grading to yellow or beige belly. Flanks covered with fine, irregular pattern of lighter spots. An offshore fish, mostly caught in waters 300 ft. (91 m) deep or more. **Range:** Nova Scotia to Florida and eastern Gulf of Mexico. **Size:** To 42 in. (1.1 m).

SAND TILEFISH
Malacanthus plumieri

Pale, very elongated body, usually yellow-brown above grading to silvery white below. Large crescent-shaped tail fin. Long dorsal and anal fins. Favors sandy areas in deeper waters and edges of coral reefs. **Range:** South Carolina to Florida, Gulf Coast, Bahamas, and Caribbean. **Size:** To 24 in. (60 cm).

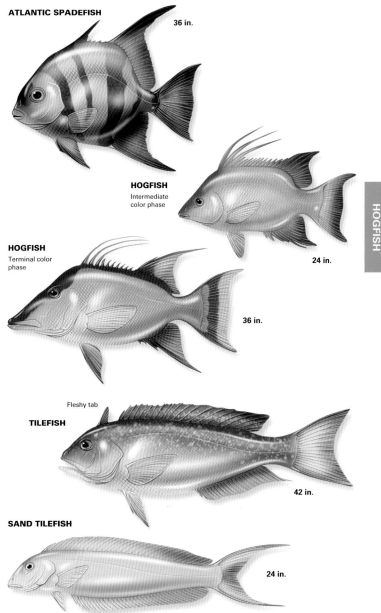

ATLANTIC SPADEFISH

36 in.

HOGFISH

Intermediate
color phase

HOGFISH

Terminal color
phase

24 in.

36 in.

Fleshy tab

TILEFISH

42 in.

SAND TILEFISH

24 in.

BASSLETS, SMALL WRASSES

FAIRY BASSLET (ROYAL GRAMMA)   *Gramma loreto*

Brightly bicolored, with deep red-violet forward and bright yellow behind. Dark spot on dorsal fin. Unusual in Florida Keys, common in Caribbean. **Range:** Florida Keys, Bahamas, and Caribbean. **Size:** To 3 in. (8 cm).

PEPPERMINT BASSLET   *Liopropoma rubre*

Mustard yellow with dark red horizontal stripes. Common on coral reefs, but very shy and hides on approach. **Range:** South Florida, Bahamas, and Caribbean. **Size:** To 3.5 in. (9 cm).

CANDY BASSLET   *Liopropoma carmabi*

Red-orange with blue horizontal stripes. Note dark spots at trailing edges of dorsal and tail fins. Common, but very shy and hides in coral caves. **Range:** Florida Keys, Bahamas, and Caribbean. **Size:** To 2 in. (5 cm).

HARLEQUIN BASS   *Serranus tigrinus*

Tiny, long-snouted, with complex pattern of dark bars or splotches along yellow to beige flanks. Common on coral reefs and sea grass beds. **Range:** South Florida, Bahamas, and Caribbean. **Size:** To 4 in. (10 cm).

TOBACCOFISH   *Serranus tabacarius*

Body variably shaded orange to white, marked with distinct dark brown patches. U-shaped dark patch at base of tail. Common on reefs. **Range:** South Florida, Bahamas, and Caribbean. **Size:** To 6 in. (15 cm).

SLIPPERY DICK   *Halichoeres bivittatus*

An elongated green, yellow, and red fish with complex patterns of spots along its flanks in the adult terminal phase. Intermediate phase is similar but much less colorful, with two horizontal stripes along flanks. **Range:** North Carolina to Florida, Bahamas, and Caribbean. **Size:** To 8 in. (20 cm).

BLUEHEAD   *Thalassoma bifasciatum*

In adult terminal phase, bright blue head, dark gill area split by a white stripe, green sides. Intermediate phase is light brown with side bars. Common on reefs. **Range:** Florida, Bahamas, southern Gulf of Mexico, and Caribbean. **Size:** To 7 in. (18 cm).

CREOLE WRASSE   *Clepticus parrae*

Adult terminal phase bicolored, with dark blue or violet forward, bright yellow to rear of flanks, and violet tail. Common on reefs. **Range:** North Carolina to Florida, Bahamas, and Caribbean. **Size:** To 12 in. (30 cm).

YELLOWCHEEK WRASSE   *Halichoeres cyanocephalus*

Elongated body, adult terminal phase with bright yellow back and broad, dark horizontal stripe along flanks. **Range:** South Florida, Bahamas, and Caribbean. **Size:** To 12 in. (30 cm).

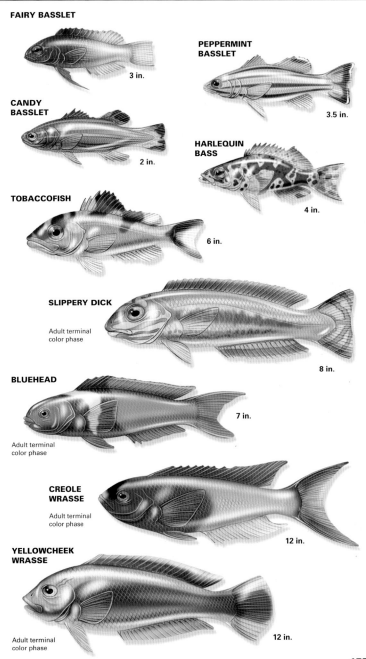

FAIRY BASSLET

3 in.

PEPPERMINT BASSLET

3.5 in.

CANDY BASSLET

2 in.

HARLEQUIN BASS

4 in.

TOBACCOFISH

6 in.

SLIPPERY DICK

Adult terminal color phase

8 in.

BLUEHEAD

Adult terminal color phase

7 in.

CREOLE WRASSE

Adult terminal color phase

12 in.

YELLOWCHEEK WRASSE

Adult terminal color phase

12 in.

PARROTFISH

STOPLIGHT PARROTFISH
Sparisoma viride

Like most parrotfish, the Stoplight changes coloration and even sex with age. Initial phase males and mature females have brick-red bellies and black-and-white scaly backs. Mature males are bright aqua-green, with blue tails and red dorsal fins. **Range:** Florida, Bahamas, and Texas Gulf Coast. **Size:** To 20 in. (51 cm).

PUDDINGWIFE
Halichoeres radiatus

Large wrasse with general looks and habits of a parrotfish. Adult terminal phase is a duller version of the intermediate phase shown. Younger fish and females are similar to intermediate phase, with brighter yellow patterning along flanks. **Range:** Georgia to Florida, Bahamas, and Gulf Coast. **Size:** To 20 in. (51 cm).

QUEEN PARROTFISH
Scarus vetula

Dull green with dramatic bright blue fin stripes and facial markings in terminal male phase. Initial and intermediate phases (juvenile and female) are dull brown or brown-gray, with gray, unmarked face and diffuse white stripes on flanks. **Range:** Florida, Bahamas, and Texas Gulf Coast. **Size:** To 24 in. (61 cm).

BLUE PARROTFISH
Scarus coeruleus

Adult terminal phase is soft powder or gray-blue, with unusual knobbed or squared-off head profile. Initial and intermediate phases are powder blue with yellow face and back fading into blue flanks. **Range:** Maryland to Florida Keys and Bahamas, absent from Gulf of Mexico. **Size:** To 30 in. (76 cm).

RAINBOW PARROTFISH
Scarus guacamaia

The largest parrotfish in our area. Terminal phase is roughly bicolored, with mustard yellow face and anterior body, and posterior flanks dark green with lighter edges to scales. Forehead gray. Initial and intermediate phases have yellow faces shading into bright green body. **Range:** Florida, Bahamas, and southern Gulf Coast. **Size:** To 36 in. (91 cm).

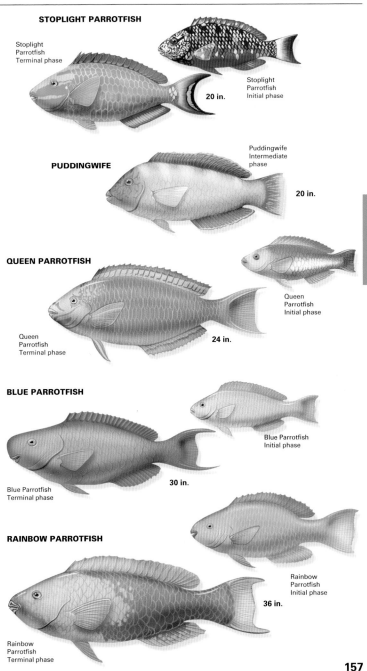

STOPLIGHT PARROTFISH

Stoplight
Parrotfish
Terminal phase

Stoplight
Parrotfish
Initial phase

20 in.

PUDDINGWIFE

Puddingwife
Intermediate
phase

20 in.

QUEEN PARROTFISH

Queen
Parrotfish
Initial phase

Queen
Parrotfish
Terminal phase

24 in.

BLUE PARROTFISH

Blue Parrotfish
Initial phase

Blue Parrotfish
Terminal phase

30 in.

RAINBOW PARROTFISH

Rainbow
Parrotfish
Initial phase

Rainbow
Parrotfish
Terminal phase

36 in.

MACKEREL

WAHOO
Acanthocybium solandri

The largest mackerel. Elongated body, with head tapering into beaklike shape. Distinct creamy vertical stripes wrap around its body. Its tenacity and fight make the Wahoo an esteemed game fish. May weigh up to 100 lbs. (45 kg). **Range:** Cape Cod south, Gulf of Mexico, and Caribbean. **Size:** To 7 ft. (2.1 m).

KING MACKEREL
Scomberomorus cavalla

Large, voracious. Spectacular royal blue above and silver below. Very swift swimmer. Comes inshore to hunt in bays. Excellent eating, but flesh spoils quickly, causing scombroid food poisoning. **Range:** Mainly from South Carolina and south, Gulf Coast, and Bahamas. **Size:** To 5.5 ft. (1.7 m).

CERO
Scomberomorus regalis

Similar to the Spanish Mackerel, but with yellow pectoral and anal fins. Flank streaked with yellow-brown spots above and below lateral line. Scales cover body, including pectoral fins. Solitary. **Range:** Mainly from South Carolina and south, Gulf Coast, Bahamas, and Caribbean. **Size:** To 3.5 ft. (1 m).

SPANISH MACKEREL
Scomberomorus maculatus

Bluish green to blue, with distinct yellowish brown to gold midflank spotting. Common game fish that schools near shores feeding on baitfish. Sensitive to water turbidity and pollution. **Range:** Massachusetts to Mexico. **Size:** To 3 ft. (91 cm).

ATLANTIC MACKEREL
Scomber scombrus

Very dark-backed with silver underparts. Occurs in large schools. Back marked with wavy bands; whitish belly. No spotting on sides. Important commercial fish, common inshore during summer. **Range:** Labrador to Cape Hatteras. **Size:** To 22 in. (56 cm).

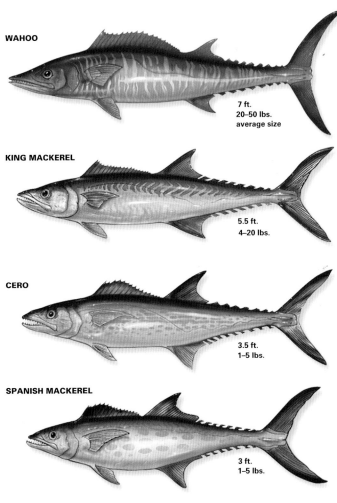

WAHOO

7 ft.
20–50 lbs.
average size

KING MACKEREL

5.5 ft.
4–20 lbs.

CERO

3.5 ft.
1–5 lbs.

SPANISH MACKEREL

3 ft.
1–5 lbs.

ATLANTIC MACKEREL

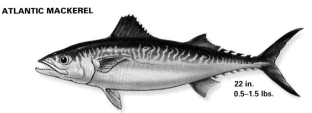

22 in.
0.5–1.5 lbs.

SMALL TUNA

Small to medium tuna species. Most are swift, offshore predators and support large sport and commercial fisheries in the Western Atlantic.

BULLET MACKEREL   *Auxis rochei*

More tuna-shaped than elongate like other mackerel, with beautiful purple tint to back and nearly vertical barring. Unspotted sides distinguish it from near look-alike Frigate Mackerel. **Range:** Bay of Fundy to Florida. **Weight:** To 15 lbs. (7 kg). **Size:** To 20 in. (51 cm).

FRIGATE MACKEREL   *Auxis thazard*

Very similar to the Bullet Mackerel but less purple tint and more oblique lines on back. One to five distinct spots below pectoral fin. **Range:** Mainly tropical waters; sometimes wanders north in Gulf Stream to offshore New York. **Weight:** To 15 lbs. (7 kg). **Size:** To 20 in. (51 cm).

LITTLE TUNNY   *Euthynnus alletteratus*

Irregular jigsaw pattern on back, numerous spots on sides and below pectoral fins. Occurs in large schools, perhaps the most common tuna off northeast coast. **Range:** Massachusetts to Brazil. **Weight:** To 25 lbs. (11 kg). **Size:** To 39 in. (1 m).

BLACKFIN TUNA   *Thunnus atlanticus*

Long, thin pectoral fin. Dusky second dorsal fin. All finlets dusky with white edges. Broad yellow-brown stripe on upper body. **Range:** Massachusetts coast to Brazil. **Weight:** To 40 lbs. (18 kg). **Size:** To 39 in. (1 m).

ATLANTIC BONITO   *Sarda sarda*

Bluish above fading to white below with distinct broad greenish stripes on back. An important commercial fish but cannot be sold as "tuna." **Range:** Migratory. Gulf of St. Lawrence to Argentina, preferring more temperate waters. **Weight:** To 20 lbs. (9 kg). **Size:** To 36 in. (91 cm).

SKIPJACK TUNA   *Katsuwonus pelamis*

Easily identified by three to five distinct black stripes on side and belly. Also note slight connection between dorsal fins. Often found in large schools. An important commercial fish. **Range:** Migratory; Nova Scotia and Newfoundland in summer, to Florida and Gulf of Mexico. **Weight:** To 75 lbs. (34 kg). **Size:** To 40 in. (102 cm).

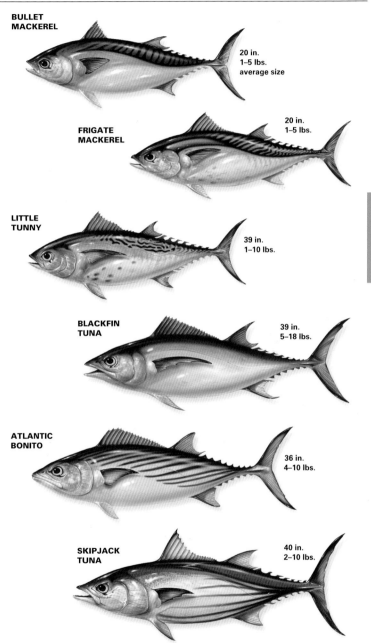

BULLET MACKEREL

20 in.
1–5 lbs.
average size

FRIGATE MACKEREL

20 in.
1–5 lbs.

LITTLE TUNNY

39 in.
1–10 lbs.

BLACKFIN TUNA

39 in.
5–18 lbs.

ATLANTIC BONITO

36 in.
4–10 lbs.

SKIPJACK TUNA

40 in.
2–10 lbs.

SMALL TUNA

LARGE TUNA

Large, swift ocean predators, highly prized for their delicious flesh. Most larger tuna species are declining in the western North Atlantic, due to overfishing by commercial fleets.

ALBACORE

Thunnus alalunga

Dark blue back, silver below. Extremely long pectoral fins. Caudal fin silver with white edge. Finlets yellow. Hunts in large, voracious schools; feeding activity often visible from a distance. **Range:** South Atlantic ranging north regularly to Long Island, rarely to Nova Scotia. **Weight:** To 100 lbs. (45 kg), averaging 20 lbs. (9 kg). **Size:** To 4.5 ft. (1.3 m); most smaller. **Red List – Data deficient**

YELLOWFIN TUNA

Thunnus albacares

Beautiful deep blue back contrasting to yellow underparts. Second dorsal very long, arched, and rich yellow. Feeds in large schools. **Range:** Nova Scotia to tropics. **Weight:** To over 400 lbs. (181 kg), averaging 20–100 lbs. (9–45 kg). **Size:** To 6 ft. (1.8 m); most smaller in our area.

BIGEYE TUNA

Thunnus obesus

Bluish above, silver below. Note short, blunt head, and very large eye. Finlets yellow with black edges. **Range:** Nova Scotia south; much more common in tropical waters. **Weight:** To 450 lbs. (204 kg). **Size:** To 7.5 ft. (2.3 m), but usually much smaller. **Red List – Vulnerable**

BLUEFIN TUNA

Thunnus thynnus

Streamlined body, dark back grading to silver on sides, creamy white below. Finlets typically gray to steel blue. Can reach huge size, although typical sizes are now much reduced from historic maximums. **Range:** Nova Scotia south. **Weight:** To 500 lbs. (227 kg). **Size:** To 14 ft. (4.3 m), but usually much smaller, averaging 7.5 ft (2.3 m). **Red List – Data deficient**

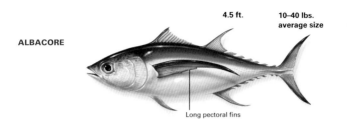

ALBACORE

4.5 ft.

10–40 lbs.
average size

Long pectoral fins

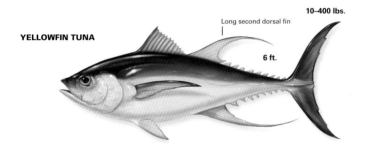

YELLOWFIN TUNA

Long second dorsal fin

10–400 lbs.

6 ft.

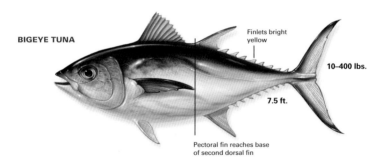

BIGEYE TUNA

Finlets bright
yellow

10–400 lbs.

7.5 ft.

Pectoral fin reaches base
of second dorsal fin

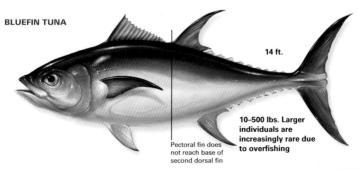

BLUEFIN TUNA

14 ft.

10–500 lbs. Larger
individuals are
increasingly rare due
to overfishing

Pectoral fin does
not reach base of
second dorsal fin

BILLFISH

BLUE MARLIN
Makaira nigricans

Massive, with elongate rostrum. Deep blue above fading to white below. Sharp, pointed, low dorsal fin. Complex lateral line, forming netlike pattern on flank. **Range:** Gulf of Maine south; prefers Gulf Stream. **Weight:** Typically 100–300 lbs. (45–136 kg). **Size:** To 15.5 ft. (4.7 m); most smaller.

WHITE MARLIN
Tetrapturus albidus

Large, blue above, white below. Dorsal fin has high, rounded front, higher than depth of body. Anal fin also rounded. Both fins may be spotted. **Range:** Gulf of Maine south in temperate waters. **Weight:** To 180 lbs. (82 kg). **Size:** To 10 ft. (3 m).

LONGBILL SPEARFISH
Tetrapturus pfluegeri

The smallest billfish. Similar to the Sailfish, but with smaller, rounded, unspotted sail. No "forehead" hump before dorsal fin, as in the Sailfish. **Range:** New Jersey south. **Weight:** To 90 lbs. (41 kg). **Size:** To 6 ft. (1.8 m).

SAILFISH
Istiophorus platypterus

Blue above, white below. Dorsal fin high, sail-like, heavily spotted. Hump on "forehead" before dorsal fin. Very small pectoral fins. Pelvic fins thin and elongate. **Range:** Tropical waters; north in Gulf Stream to New York. **Weight:** To 180 lbs. (82 kg). **Size:** To 11 ft. (3.4 m).

SWORDFISH
Xiphias gladius

The largest billfish. Grayish, greenish, or blue with dark fins. Very long rostrum ("sword"). Long peduncle keel forward and on side of caudal fins. Frequents the water's surface, where it is often harpooned. **Range:** Grand Banks to tropics. **Weight:** To 1,300 lbs. (590 kg). **Size:** To 15 ft. (4.6 m). **Red List – Data deficient**

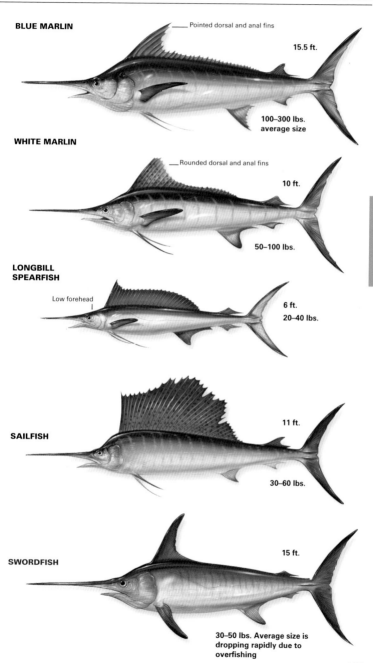

BLUE MARLIN

Pointed dorsal and anal fins

15.5 ft.

100–300 lbs. average size

WHITE MARLIN

Rounded dorsal and anal fins

10 ft.

50–100 lbs.

LONGBILL SPEARFISH

Low forehead

6 ft. 20–40 lbs.

SAILFISH

11 ft.

30–60 lbs.

SWORDFISH

15 ft.

30–50 lbs. Average size is dropping rapidly due to overfishing

BILLFISH

LEFTEYE FLOUNDERS

GULF STREAM FLOUNDER
Citharichthys arctifrons

Rather plain with no distinctive or characteristic coloration. Small, brown to light brown, with small, hornlike projection on snout. **Range:** Cape Cod to southern Florida and Gulf Coast. **Size:** To 7 in. (18 cm).

SUMMER FLOUNDER
Paralichthys dentatus

Dark brown or gray-brown with many eyespots (ocelli) on the side, although color is highly variable, based on the local environment. Favors sandy or muddy bottoms at shallow to moderate depths, often near marine grass beds. **Range:** Maine to northern Florida. **Size:** To 36 in. (91 cm).

SOUTHERN FLOUNDER
Paralichthys lethostigma

No eyespots (ocelli) on the side, but often with irregular or diffuse darker patches. Exact color varies with local habitat, but most are medium to dark brown. **Range:** North Carolina and Gulf Coast, but absent from southern Florida. **Size:** To 33 in. (84 cm).

GULF FLOUNDER
Paralichthys albigutta

Very similar to the Southern Flounder, but smaller and almost always with three distinct dark eyespots (ocelli) on the left ("top") side. Favors flat sandy bottoms near sea grass beds, rocky reefs, and piers and other structures. **Range:** North Carolina to Florida, Bahamas, and Gulf Coast. **Size:** To 16 in. (41 cm).

WINDOWPANE
Scophthalmus aquosus

Common but fairly nondescript. Color is highly variable based on local habitat, but always with a fine pattern of mottling and scattered darker blotches. First few rays of dorsal fin near eyes are long and branched. **Range:** Gulf of St. Lawrence to northern Florida. **Size:** To 18 in. (46 cm).

GULF STREAM FLOUNDER
Small "horn" on snout

SUMMER FLOUNDER

7 in.

36 in.

SOUTHERN FLOUNDER

33 in.

GULF FLOUNDER

WINDOWPANE
Long first rays of dorsal fin

18 in.

16 in.

Three dark eyespots

FLOUNDERS

BAY WHIFF (DAB)
Citharichthys spilopterus

Small but otherwise very nondescript lefteye flounder that favors muddy bottoms of coastal bays and shallow flats. Color varies from medium to dark brown, with fine speckle patterns of darker blotches. No eyespots (ocelli) or other large spots. **Range:** New Jersey to Florida and northern Gulf Coast. **Size:** To 10 in. (20 cm), but averages 6 in. (15 cm).

WINTER FLOUNDER
Pseudopleuronectes americanus

Also known as the Flatfish, this species is well known to sportfishers. Mainly a shallow-water species that ranges from Labrador to Chesapeake Bay. Larger individuals are found in deeper water in excess of 2,100 ft. (640 m). Winter Flounders migrate from deep offshore waters to shallow coastal waters in fall, and move back to deep waters in spring. Larger individuals can be nearly 2 ft. (61 cm) long and weigh up to 5 lbs. (2 kg). Winter Flounders caught off Rhode Island are often quite large and are called Snowshoes by locals. **Range:** Labrador to Georgia. **Size:** Typically 15 in. (38 cm), but may reach 24 in. (61 cm).

ATLANTIC HALIBUT
Hippoglossus hippoglossus

Former mainstay of East Coast fishing industry, now overfished throughout its range. Large, bottom-feeding flounder that eats other fish, including cod, herring, and skates. Halibut in turn is food for seals and is taken by Greenland Sharks in northern waters. **Range:** Formerly Greenland to Cape Hatteras coast in cooler offshore waters; now mostly restricted to areas north of Delmarva Peninsula. **Size:** Formerly to 6 ft. (1.8 m) or more, now typically 3-4 ft. (0.9-1.2 m) or less. **Red List – Endangered**

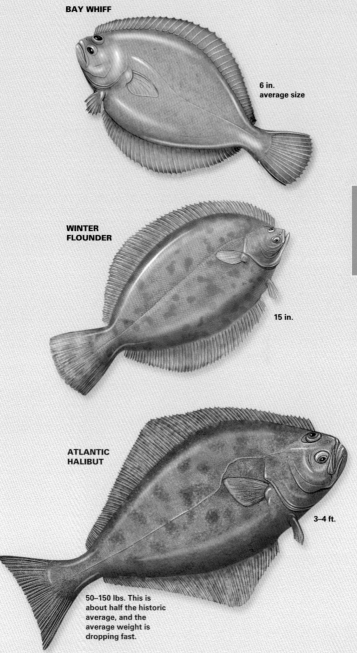

BAY WHIFF

6 in.
average size

WINTER FLOUNDER

15 in.

ATLANTIC HALIBUT

3–4 ft.

50–150 lbs. This is about half the historic average, and the average weight is dropping fast.

SOLES

NAKED SOLE
Gymnachirus melas

A right-eyed flatfish, medium brown with 20–30 dark thick bars across upper side. Frequents sandy and muddy bottoms of deeper coastal waters. **Range:** Cape Cod to Florida, Bahamas. **Size:** To 9 in. (23 cm).

FRINGED SOLE
Gymnachirus texae

Smaller and lighter-colored than the Naked Sole, with more and finer barring on upper side. Almost circular body outline. Frequents sandy and muddy silt bottoms, often near cover of rocks, reefs, or grass flats. **Range:** North Carolina to Florida and Gulf Coast to Yucatán. **Size:** To 6 in. (15 cm).

SCRAWLED SOLE
Trinectes inscriptus

Light brown or tan body covered with a fine netting pattern of dark lines. Found in clear salt water on flat, sandy bottoms and near mangroves. **Range:** South Florida and Bahamas, but absent from Gulf of Mexico. **Size:** To 6 in. (15 cm).

HOGCHOKER
Trinectes maculatus

Almost circular shape. Color highly variable, with some fish pale to dark gray and others, more numerous, in shades from light to dark brown. Favors shallow coastal waters and brackish bays, and enters freshwater rivers for miles upstream. **Range:** Massachusetts to Florida and eastern Gulf Coast. **Size:** To 8 in. (20 cm).

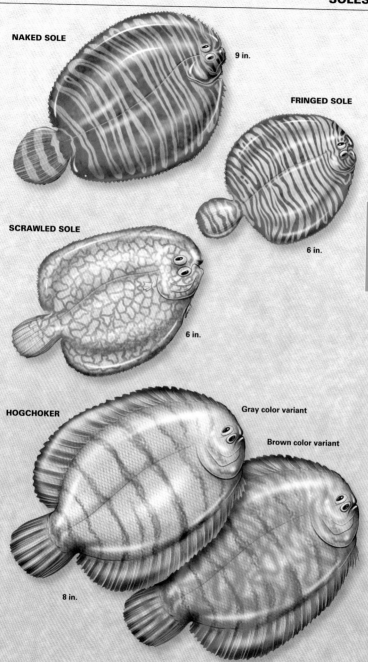

NAKED SOLE

9 in.

FRINGED SOLE

6 in.

SCRAWLED SOLE

6 in.

HOGCHOKER

Gray color variant

Brown color variant

8 in.

SMOOTH TRUNKFISH
Lactophrys triqueter

Characteristically odd trunkfish body shape, dark gray with fine spattering of white spots. Honeycomb pattern of lines in middle of flank, behind pectoral fin. Found above and around coral reefs and rocky reefs offering shelter. **Range:** Massachusetts to Florida, Bahamas, and Gulf of Mexico. **Size:** To 12 in. (30 cm).

SCRAWLED COWFISH
Acanthostracion quadricornis

Lemon to mustard yellow body covered with pattern of irregularly shaped blue lines and blotches. Found in a range of habitats, from coral reefs and sea grass beds to deeper rock reefs. **Range:** Cape Cod to Florida, Bahamas, Gulf Coast, and Caribbean. **Size:** To 19 in. (48 cm).

SPOTTED TRUNKFISH
Lactophrys bicaudalis

Yellow to green-yellow or tan body with an even pattern of dark spots, finer on head and larger on flanks. A wary inhabitant of coral reefs, where it usually hovers near the shelter of small caves and hollows. **Range:** Florida, Bahamas, and Texas Gulf Coast. **Size:** To 19 in. (48 cm).

TRUNKFISH
Lactophrys trigonus

The classic trunkfish profile, but in an even, dull green or aqua-green, unmarked by distinctive patterns of spots. Darker individuals are sometimes olive green to brown. **Range:** Cape Cod to Florida, Bahamas, Gulf Coast, and Caribbean. **Size:** To 18 in. (46 cm).

HONEYCOMB COWFISH
Acanthostracion polygonius

Dramatic and distinctive pattern honeycomb hex pattern of dark lines covers pale tan or yellow body. Called "cowfish" for sharp spines above each eye. A shy inhabitant of coral reefs and other rocky reef areas that offer shelter. **Range:** New Jersey to Florida, Bahamas, and Caribbean, but absent from Gulf of Mexico. **Size:** To 18 in. (46 cm).

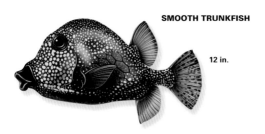

SMOOTH TRUNKFISH

12 in.

SCRAWLED COWFISH

19 in.

SPOTTED TRUNKFISH

19 in.

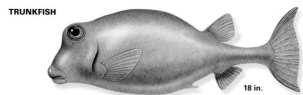

TRUNKFISH

18 in.

HONEYCOMB COWFISH

18 in.

FILEFISH

SCRAWLED FILEFISH   *Aluterus scriptus*

Mustard yellow to lime green body covered with a complex pattern of blue lines and darker spots. Common if wary resident of coral reefs, where it skulks under the shelter of soft corals and near overhangs in reefs. **Range:** Nova Scotia to Florida, Bahamas, and Caribbean. **Size:** To 30 in. (76 cm).

WHITESPOTTED FILEFISH   *Cantherhines macrocerus*

Oddly shaped body, with an extended downward projection of belly. Roughly bicolored, with gray head and back shading to mustard or greenish yellow belly and sides. Covered with heavy pattern of light spots. Common variant has same colors but no spots. Found on coral reefs. **Range:** Cape Canaveral to south Florida, Bahamas, and Caribbean. **Size:** To 18 in. (46 cm).

UNICORN FILEFISH   *Aluterus monoceros*

Distinct profile, blade-shaped with a concave indentation below a relatively small mouth. Uniform blue-gray or gray-brown. Generally found in open water over or around reefs. **Range:** Cape Cod to Florida and Bahamas, uncommon in Gulf of Mexico. **Size:** To 24 in. (61 cm).

ORANGE FILEFISH   *Aluterus schoepfii*

Colors and patterns vary widely in this species. Gray to brown head and back, shading to orange or red-orange across sides. Some are even a silvery gray. Body may be relatively unmarked or heavily marked with darker spots and blotches. **Range:** Nova Scotia to Florida, Bahamas, and Gulf Coast. **Size:** To 24 in. (61 cm).

PLANEHEAD FILEFISH   *Monacanthus hispidus*

Muted tan or gray-brown body covered with small dark spots. Deep downward projection of the belly profile. Found near the bottom over rocky reefs, coral reefs, sea grass beds, and other sheltered areas. **Range:** Nova Scotia to Florida, Bahamas, Gulf Coast, and Caribbean. **Size:** To 10 in. (25 cm).

ORANGESPOTTED FILEFISH   *Cantherhines pullus*

Color variable, but usually dark with four or five horizontal rows of fine orange or yellow spots. Common over rocky reefs, coral reefs, and other bottoms offering shelter. **Range:** Cape Cod to Florida, Bahamas, Gulf Coast, and Caribbean. **Size:** To 8 in. (20 cm).

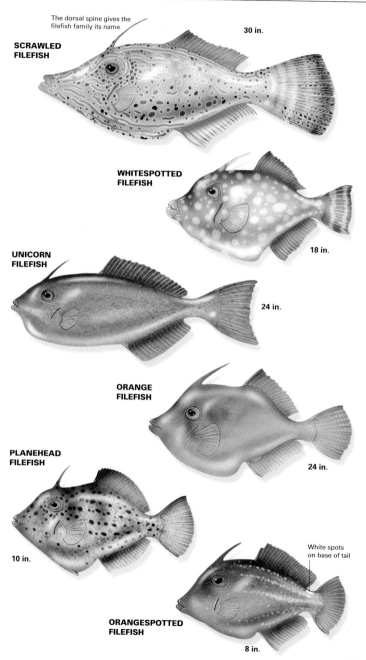

The dorsal spine gives the filefish family its name

30 in.

SCRAWLED FILEFISH

WHITESPOTTED FILEFISH

18 in.

UNICORN FILEFISH

24 in.

ORANGE FILEFISH

PLANEHEAD FILEFISH

24 in.

10 in.

White spots on base of tail

ORANGESPOTTED FILEFISH

8 in.

FILEFISH

175

TRIGGERFISH

SARGASSUM TRIGGERFISH
Xanthichthys ringens

Highly variable body color and can change color at will. Mostly tan to yellow, with lines of fine spots across flanks. Three lines below eye. Young live in *Sargassum* weed mats, adults near the bottom in deeper water. **Range:** Cape Cod to Florida, Bahamas, Gulf of Mexico, and Caribbean. **Size:** To 10 in. (25 cm).

GRAY TRIGGERFISH
Balistes capriscus

Even olive gray to silver-gray, with dark vertical stripes across flanks. Up close, note subtle pattern of fine blue lines and spots. Near mats of *Sargassum* weed at the surface or near the bottom in small schools. **Range:** Nova Scotia to Florida, Bahamas, Gulf of Mexico, and Caribbean. **Size:** To 12 in. (30 cm).

OCEAN TRIGGERFISH
Canthidermis maculata

Uniform blue-gray to silver-gray above shading to white belly. Dark spot near base of pectoral fin. Strong swimmer, either alone or in small groups well above the bottom in open water. **Range:** Cape Cod to Florida, Bahamas, Gulf of Mexico, and Caribbean. **Size:** To 24 in. (61 cm).

BLACK DURGON
Melichthys niger

Very dark blue, with fine pattern of black scales throughout. Bright blue lines around eyes and at base of dorsal and anal fins. Usually seen in small groups swimming above coral reefs. **Range:** Florida, Bahamas, Gulf of Mexico, and Caribbean. **Size:** To 10 in. (25 cm).

QUEEN TRIGGERFISH
Balistes vetula

Dramatic yellow and green body, with fine patterns of bright blue lines and spots. Blue fins. More common in warmer waters, where it is more brightly colored than in the northern part of its range. Found over reefs and sea grass beds. **Range:** Cape Cod to Florida, Bahamas, Gulf of Mexico, and Caribbean. **Size:** To 24 in. (61 cm).

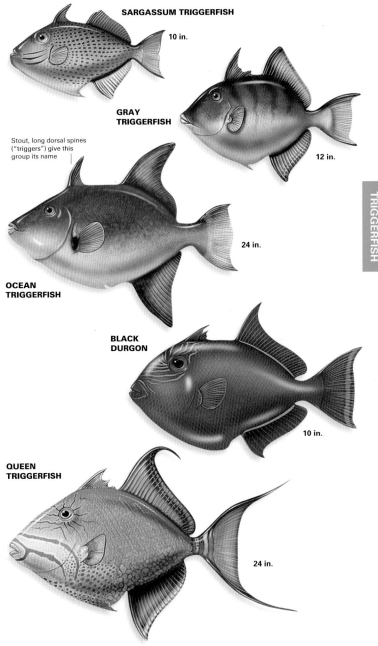

SARGASSUM TRIGGERFISH

10 in.

GRAY TRIGGERFISH

Stout, long dorsal spines ("triggers") give this group its name

12 in.

24 in.

OCEAN TRIGGERFISH

BLACK DURGON

10 in.

QUEEN TRIGGERFISH

24 in.

TRIGGERFISH

177

PUFFERS, PORCUPINEFISH

These fish are named for their unusual habit of swallowing water or air to increase their size, thereby detering predators from swallowing them. Most puffers cannot be eaten owing to the presence of a neurotoxin known as tetrodotoxin in their flesh.

NORTHERN PUFFER   *Sphoeroides maculatus*

Olive gray with many black dots speckled over upper side. Black band between eyes. Fins usually yellowish, with yellow edging on dark side stripes. A fish of inshore waters, harbors, and estuaries. Nontoxic and sold as "sea squab." **Range:** Cape Cod to Florida. **Size:** To 14 in. (36 cm).

SMOOTH PUFFER   *Lagocephalus laevigatus*

Bluish gray overall, with indistinct broad banding (bands are more distinct in young specimens). No spines. Forked caudal fin. Mainly pelagic. Adults are seen at the ocean's surface far offshore. Young are more common inshore on coastal banks. **Range:** Cape Cod to Argentina. **Size:** To 24 in. (61 cm).

OCEANIC PUFFER   *Lagocephalus lagocephalus*

Similar to the Smooth and Northern Puffers, but darker blue above, with contrasting white undersurface. When viewed floating in surface waters, note how close dorsal fin is to caudal. As its name suggests, this is the most pelagic of the puffers, and healthy adults are rarely encountered in inshore waters. **Range:** Nova Scotia to Florida. **Size:** To 24 in. (61 cm).

PORCUPINEFISH   *Diodon hystrix*

Yellowish tan body covered with spines that can be elevated. Numerous small spots on body. Large eyes. Similar to the smaller Balloonfish (*D. holocanthus*), but spines on forehead are shorter than those on body. **Range:** Worldwide in warm waters, drifts north along the Atlantic Coast to Cape Cod. **Size:** To 36 in. (91 cm).

SPOTTED BURRFISH   *Chilomycterus atringa*

Similar to a small Porcupinefish. Yellowish tan body with very short spines that appear triangular in cross-section. Note distinct black spotting on back and fine spotting covering all fins. **Range:** New Jersey south to Caribbean and Gulf of Mexico. **Size:** To 18 in. (46 cm).

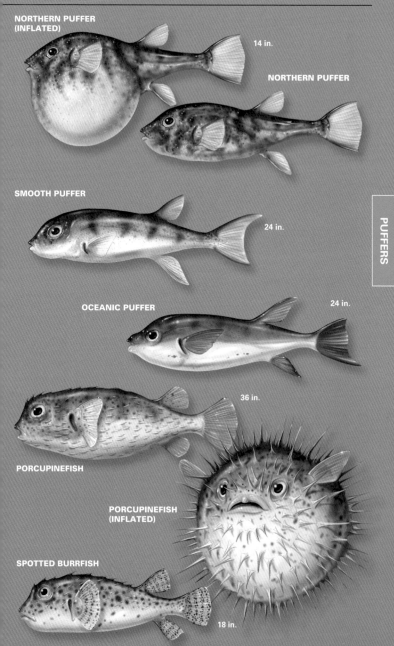

NORTHERN PUFFER (INFLATED)

14 in.

NORTHERN PUFFER

SMOOTH PUFFER

24 in.

OCEANIC PUFFER

24 in.

36 in.

PORCUPINEFISH

PORCUPINEFISH (INFLATED)

SPOTTED BURRFISH

18 in.

OCEAN SUNFISH   *Masturus lanceolatus, Mola mola*

These pelagic giants are unmistakable: massive, with a flattened, circular body that looks as if it had been chopped off at the rear. The mouth is small, and the gill opening is a mere hole. All ocean sunfish, or molas, lack pectoral fins, and the dorsal and anal fins have exceptionally long rays, giving them a sailboatlike appearance from the side. When floating near the water's surface they are often accompanied by seabirds picking at parasites found in the thick mucus covering of the skin. Ocean sunfish feed mainly on oceanic jellyfish like the Portuguese Man-of-War, other invertebrates, and some larval fish. These species are worldwide in distribution in tropical waters.

SHARPTAIL MOLA   *Masturus lanceolatus*

Massive, very similar to *Mola mola*. Caudal "fin" shows a distinct central projection, hence its name. Though often seen floating listlessly at the water's surface with dorsal fin projecting into the air, this behavior is not typical of healthy adults, and molas who surface may be injured or dying. Despite their large size, molas are harmless unless attacked. Not considered edible. **Weight:** To 2 tons (1.8 t) or more. **Size:** To 10 ft. (3 m) long and 11 ft. (3.3 m) high.

OCEAN SUNFISH   *Mola mola*

Massive disklike body, brownish olive to slate gray. Mouth tiny and round. Elongate vertical rays in tall dorsal and anal fins are used to scull water in tropical areas, where it can swim remarkably fast. Dorsal fin often projects from the water's surface. Caudal "fin" is merely a wavy flap of uniform width, with no projection as in the Sharptail Mola. **Weight:** To 2 tons (1.8 t) or more. **Size:** To 10 ft. (3 m) long and 11 ft. (3.3 m) high.

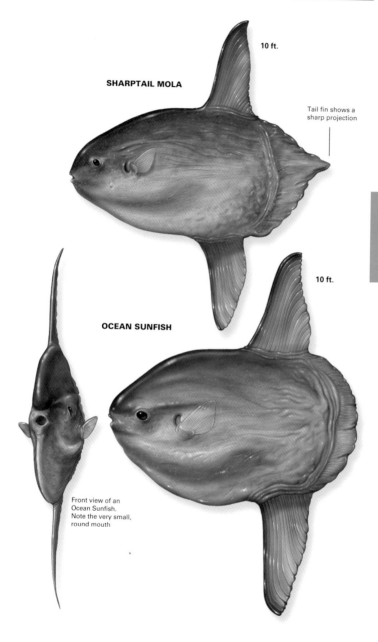

SHARPTAIL MOLA

10 ft.

Tail fin shows a
sharp projection

10 ft.

OCEAN SUNFISH

Front view of an
Ocean Sunfish.
Note the very small,
round mouth

GREEN TURTLE, HAWKSBILL TURTLE

Sea turtles are mainly inhabitants of tropical waters. The smaller turtle species are sometimes difficult to identify when seen at the surface. They are often covered with barnacles and may be in poor physical condition with battered shells. In some cases only a careful examination of the shell scutes (bony plates) will confirm an identification. Beware of handling these large, powerful animals. *Sea turtles are protected animals in all US waters* and should not be handled unless they are fouled in fishing gear, and then only with great caution. Contact a marine animal rescue unit rather than attempting to save a turtle yourself.

GREEN TURTLE
Chelonia mydas

Olive brown overall; the "green" of its name refers to the color of its body fat. Shell is flecked and mottled with lighter patterns. Large head is distinctly scaled, with pale margins on scales. Front flippers are relatively large, with a single claw on each. The serrated lower jaw edges are unique to the Green Turtle. May swim at the water's surface with a bobbing motion, its back alternately surfacing and dipping below water. Feeds principally on sea grass, crabs, jellyfish, and invertebrates. The only US nesting area is in Florida. **Range:** Cape Cod south, and throughout Caribbean. **Size:** To 5 ft. (1.5 m). **Red List – Endangered**

HAWKSBILL TURTLE
Eretmochelys imbricata

Relatively small sea turtle with a rich chestnut-brown shell with black and yellow "tortoiseshell" markings. Shell scales have ragged rear edges that slightly overlap. Elongated "hawk's bill"–like upper jaw, hence its name. Feeds principally on marine sponges. Nests on Caribbean beaches. Most common in tropical waters, especially near coral reefs. Drifts northward in Gulf Stream. Atlantic Hawksbill numbers have suffered greatly due to hunting for shells and meat. **Range:** Cape Cod south, and throughout the Caribbean. **Size:** To 3 ft. (91 cm). **Red List – Critically endangered**

GREEN TURTLE

5 ft.

Serrated edge of
lower jaw

Single claw on each
front flipper

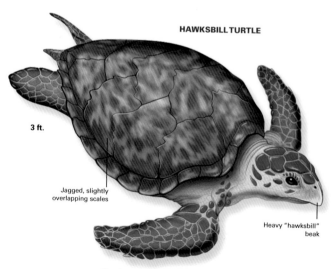

HAWKSBILL TURTLE

3 ft.

Jagged, slightly
overlapping scales

Heavy "hawksbill"
beak

Two claws on each
front flipper

SEA TURTLES

LOGGERHEAD SEA TURTLE
Caretta caretta

Bright reddish brown shell is diagnostic, as are the large head and heavy crushing jaws. Elongated shell with serrated ridges on rear edges. Large head with distinctive hump at nape. The most inshore of the sea turtles, regularly entering bays and estuaries as well as more offshore waters. Diet includes mollusks, sponges, Horseshoe Crabs, other invertebrates, Turtle Grass, and marine algae. Nests along southeast coast between April and August. Regularly nests as far north as Carolinas, rarely into Maryland. Loggerheads can live up to 20 years. **Range:** Throughout Atlantic Coast and Caribbean; commonly wanders north to Nova Scotia. **Size:** To 7 ft. (2.1 m). **Red List – Endangered**

KEMP'S RIDLEY
Lepidochelys kempi

Very small size (for a sea turtle) is diagnostic. Olive green to yellowish in color; pale older individuals can be nearly white in patches. Shell is heart-shaped with three ridges in young specimens. Lateral back ridges become much less prominent with age; adults have smooth shells. Overall shape more rounded and compact than other sea turtles. Diet includes crabs, mollusks, and fish. Nests mainly in Texas and Mexico, rarely in Florida. **Range:** Throughout the southeast Atlantic Coast and Caribbean, but never common, particularly in US waters. **Size:** To 29 in. (74 cm). **Red List – Critically endangered**

The large head and massive parrot beak of the Loggerhead Sea Turtle are distinctive.

P. LYNCH

LOGGERHEAD SEA TURTLE

7 ft.

Proportionately
large head

Two claws on each
front flipper

KEMP'S RIDLEY

29 in.

Typically olive green,
and *very small* for a
sea turtle

Single claw on each
front flipper

Distinctive features: The largest sea turtle. Bluish black to dark gray, flecked with white or pinkish spotting. The large size and the seven prominent ridges arising from the flexible "leatherback" carapace are immediately diagnostic. Note the very long front flippers. **Red List – Critically endangered**

Description and size: This huge turtle can reach 9 ft. (2.7 m) in length and more than 1,500 lbs. (680 kg) in weight, with record animals weighing in at over a ton. When viewed from above, the leathery-skinned "shell" tapers toward the rear, giving the Leatherback a heart-shaped appearance.

Range: Throughout the Atlantic Ocean and Caribbean, but never common; particularly uncommon in northern Gulf waters.

Habits: A wanderer of the deep oceans, this sea turtle is remarkably adapted to range into the cold deep North Atlantic waters, but it is also commonly found in more shallow warmer waters of the southeastern and Gulf coasts and throughout the Caribbean.

The Leatherback is a deep diver; it can dive as deep as Sperm Whales, reaching depths of several thousand feet. Its diet consists mainly of jellyfish, making this sea turtle one of the few animals to use this group as a food source. When it encounters large masses of jellyfish, it can spend hours feeding in the area. Normally solitary. Loose groups of leatherbacks can be encountered on occasion, especially during concentrated jellyfish outbreaks and near nesting beaches along the southeast Atlantic Coast in spring and summer.

Similar species: When well viewed, the flexible carapace with its prominent ridges and the turtle's overall size make this species unmistakable.

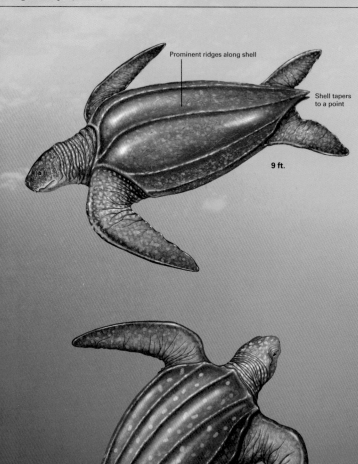

Prominent ridges along shell

Shell tapers
to a point

9 ft.

AMERICAN CROCODILE
Crocodylus acutus

Gray-green to olive green, often with dark crossbands. Head profile viewed from above is much narrower than the broad-headed American Alligator. The large fourth tooth of the lower jaw can be seen outside the mouth when it is closed, unlike the hidden fourth tooth of the alligator. Females build a mound of vegetation or sandbank cavity and lay 30–50 eggs in April–May. Hatchlings emerge in July–August. Inhabits bays, mangrove swamps, and boggy coastal sites. Aggressive protection of crocodiles and their habitat has resulted in a significant population increase. **Range:** Extreme southern Florida (Biscayne Bay, Everglades) and Florida Keys. **Size:** To 18 ft. (5.5 m) in the United States. **Red List – Vulnerable**

AMERICAN ALLIGATOR
Alligator mississippiensis

Large reptile, distinguished from the much less common American Crocodile by its broad head profile and rounded snout. Dark gray to black. Young have cream-colored crossbands that fade with age. Large lower fourth tooth does not show when the mouth is closed. The male booms out loud roars to attract the female. During dry spells, "gators" often excavate deeper holes to hold water, which keeps their skins from drying out. Found in a wide range of habitats from fresh to brackish marshes, lakes, ponds, rivers, and swamplands. **Range:** East Coast from southeastern North Carolina to the Florida Keys and entire Gulf Coast. **Size:** To 19 ft. (5.8 m). **Red List – Lower risk/least concern**

Alligator. Note the *very broad, rounded snout* and high brow ridges. Inhabits a wide range of freshwater and brackish habitats throughout the Southeast.

Crocodile. Note the *long, narrow snout* and relatively low brow ridges. More sensitive to cold weather than the alligator, which restricts its range to extreme south Florida and the Keys.

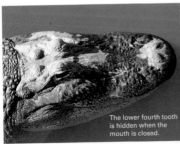

The lower fourth tooth is hidden when the mouth is closed.

The lower fourth tooth juts out when the mouth is fully closed.

P. LYNCH

AMERICAN
ALLIGATOR

AMERICAN
CROCODILE

LOONS, GREBES

Loons and grebes are long-necked diving birds most commonly seen in marine waters during spring and fall migration and when wintering in salt water. These birds rarely venture far offshore.

COMMON LOON
Gavia immer

A large, heavy-bodied bird mainly of inshore waters. Long, heavy bill is adapted to capture fish during deep dives. Seen mainly in winter and during spring and fall migration. Winter bird is dark-backed and white below. Bill held parallel to water and not uptilted. Found in winter along Atlantic and Gulf Coasts except in south Florida. **Length:** 25 in. (64 cm). **Wingspan:** 58 in. (147 cm).

RED-THROATED LOON
Gavia stellata

Distinguished from the Common Loon by its smaller size and thin, upturned bill. In winter plumage, much grayer than the Common Loon. Thin bill usually points upward. Favors inshore waters. Seen mainly in winter and during spring and fall migration along the Atlantic and northeastern Gulf Coasts. **Length:** 25 in. (64 cm). **Wingspan:** 44 in. (112 cm).

HORNED GREBE
Podiceps auritus

A very small diving bird with dark gray back and light gray to white underparts. Top of head appears black-capped. Light gray throat and cheeks. Delicate, short bill. Not a deep ocean bird, favoring inshore waters. Often seen in flocks of hundreds at harbor mouths. Found in winter throughout southeastern states and along Atlantic and Gulf Coasts; uncommon in southern Florida and southern Texas. **Length:** 13 in. (33 cm). **Wingspan:** 25 in. (64 cm).

PIED-BILLED GREBE
Podilymbus podiceps

A small, brown, short-necked diving bird that in breeding plumage has a distinct dark ring around the light bill. In contrast to the Horned Grebe, the Pied-billed shows no white areas on the wing while in flight. Breeds inland in freshwater areas, but is very common in coastal brackish and saltwater environments in winter. **Range:** Throughout the United States, and in winter especially common along the Atlantic and Gulf Coasts and Bahamas. **Length:** 13 in. (33 cm). **Wingspan:** 25 in. (64 cm).

COMMON LOON

Winter (nonbreeding)

Long, heavy bill

Head is held lower than body

Winter

RED-THROATED LOON

Thin, upturned bill

Winter

Winter

HORNED GREBE

Winter

Winter

PIED-BILLED GREBE

Winter

Breeding

Winter (nonbreeding)

FRIGATEBIRD, TROPICBIRD

WHITE-TAILED TROPICBIRD

Phaethon lepturus

Unmistakable. Beautiful white seabird with bold black chevrons on back and very long white tail streaming behind. Yellow bill. Immature has a black-barred back. Buoyant flight. Inquisitive, will join other seabirds in feeding flocks. **Length:** 31 in. (79 cm). **Wingspan:** 38 in. (97 cm).

MAGNIFICENT FRIGATEBIRD

Fregata magnificens

Long-winged but very streamlined. Note the long, forked tail. Male solid black, with a red throat pouch mostly hidden from view when in flight. Males inflate their throat pouches near the nest in a courtship display. Female has a dark head and a white underbelly. Immature has an all-white head and a white underbelly. Pursues and robs food from other seabirds. **Length:** 40 in. (102 cm). **Wingspan:** 92 in. (2.3 m).

Bird and Hospital Keys, just opposite Fort Jefferson in the Dry Tortugas National Park. The Dry Tortugas are the farthest extensions of the Florida Keys archipelago, lying about 70 miles west of Key West. These two small islands are protected nesting areas for Sooty Terns, frigatebirds, and sea turtles.

P. LYNCH

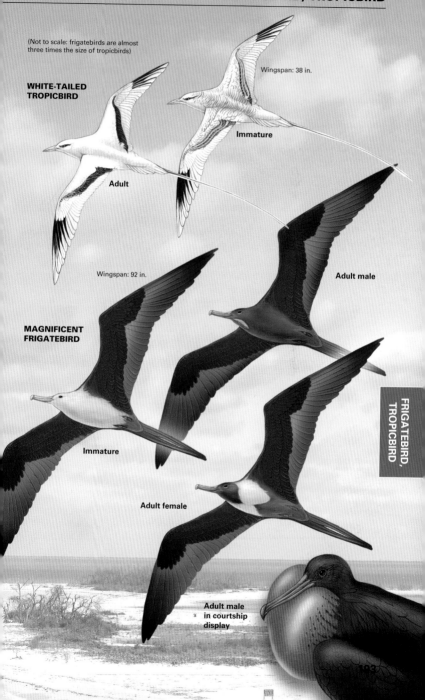

(Not to scale: frigatebirds are almost three times the size of tropicbirds)

WHITE-TAILED TROPICBIRD

Wingspan: 38 in.

Immature

Adult

Wingspan: 92 in.

Adult male

MAGNIFICENT FRIGATEBIRD

Immature

Adult female

Adult male in courtship display

Description and size: This medium-sized seabird is superficially similar to a gull. The body proportions, however, are quite different: fulmars are much bulkier and thicker-necked than gulls, and they have a high, domed forehead. The Northern Fulmar has two distinct color phases. Most fulmars in our area are light-phase birds, but dark-phase individuals also occur regularly. Intermediate-phase birds blend the characteristics of the dark and light phases. **Length:** 19 in. (48 cm). **Wingspan:** 43 in. (109 cm).

Habits: Although they look much like gulls, fulmars behave very differently, and this makes them fairly easy to distinguish from gulls. Fulmars have a distinctive, stiff-winged glide that is closer to the flight of shearwaters than it is to any gull flight pattern. Fulmars are deep-ocean wanderers that readily follow ships and take chum. They are often seen near feeding whales, looking for scraps and injured fish. The fulmar population has increased dramatically in the past two decades.

Similar species: At longer ranges, light-phase Northern Fulmars are first distinguished from gulls by their stiff-winged flight characteristics and chunkier body silhouette. Unlike gulls, fulmars are rarely seen within sight of land, except near their Arctic nesting colonies. Dark-phase Northern Fulmars are similar in color to Sooty Shearwaters but have much bulkier bodies and proportionately broader wings than the slimmer, more slender-winged shearwaters.

NORTHERN FULMAR

LIGHT PHASE

Bill is thick and short

Body is chunky and more compact than gulls

Inner primaries are light at the base

Fulmar wingbeats are stiff and shallow; gull wingbeats are deeper

DARK PHASE

Prominent forehead with thick neck

NORTHERN FULMAR

195

LARGE SHEARWATERS

Shearwaters have long, thin wings that are perfectly adapted for rapid flight over the ocean surface. Their wingtips occasionally slice into the waves of turbulent water, hence the name "shearwater." The stiff, unflexed wingbeat of shearwaters separates them from all other seabirds. Shearwaters nest in the South Atlantic Ocean during the austral summer and spend the austral winter (our summer) in North Atlantic waters.

GREATER SHEARWATER *Puffinus gravis*

Large, with a distinctive black cap delineated by a white nape. Hard separation between dark upper parts and white underbelly. Stark white rump contrasts with dark back. Note dark underbelly smudge. The most common shearwater in our offshore waters; hundreds can be seen at times. Comes readily to chum. **Length:** 18 in. (46 cm). **Wingspan:** 44 in. (112 cm).

BLACK-CAPPED PETREL *Pterodroma hasitata*

Very large, like a shearwater. Dark gray with white forehead and distinct U-shaped white rump patch. White wing linings. Note distinct bend to wing and fluttery flight pattern. Rollercoaster-like flight, soaring up well above water, then plunging down with bowed wings. Breeds on Caribbean islands. **Length:** 16 in. (41 cm). **Wingspan:** 37 in. (94 cm).

CORY'S SHEARWATER *Calonectris diomedea*

The largest of our shearwaters. Underbelly lacks dark smudge of the Greater Shearwater. Note distinct yellow bill. Flight sequence is usually four slow flaps followed by a long glide. Never common or even regular in our area, usually seen individually. Comes quite close to land and can often be seen from coastal capes and sea islands. **Length:** 20 in. (51 cm). **Wingspan:** 44 in. (112 cm).

SOOTY SHEARWATER *Puffinus griseus*

Appears as all brown, but note gray-white underwings. One of the most common shearwaters. At times, hundreds mass on the ocean surface, apparently feeding on floating krill. Readily approaches ships and will come to chum and settle on the water. **Length:** 17 in. (43 cm). **Wingspan:** 43 in. (109 cm).

GREATER SHEARWATER

Black cap and smudge on belly

Off the Carolina coast, the Black-Capped Petrel may be confused with the Greater Shearwater; the petrel has a white forehead, a much broader rump patch, and no dark smudge on the belly

BLACK-CAPPED PETREL

CORY'S SHEARWATER

Cap is smudgy and lacks contrast

SOOTY SHEARWATER

Gray wing linings

LARGE SHEARWATERS

197

SMALL SHEARWATERS

Two species of small, closely related shearwaters. Only the Manx Shearwater is commonly seen. Audubon's Shearwater is more frequent south of Virginia, but unusual compared to the Manx. The much larger Greater Shearwater is shown for comparison.

AUDUBON'S SHEARWATER   *Puffinus lherminieri*

The smallest of our regularly occurring shearwaters. Near look-alike to the Manx Shearwater but with dark undertail and distinctly longer tail. Best clue is flight pattern, very fluttery and fast as the bird rocks back and forth over the water. Single birds are usually seen and, because of their small size, can be overlooked when mixing with larger shearwater species. **Length:** 12 in. (30 cm). **Wingspan:** 27 in. (69 cm).

MANX SHEARWATER   *Puffinus puffinus*

A small black-and-white shearwater, half the size of the bulky Greater Shearwater. The Manx lacks the Greater's capped appearance and white rump. Flight is very buoyant and bounding, as if being pulled up by a string while fluttering on stiff wings. When looking at groups of Manx Shearwaters from a distance, the very strong contrast between the solid dark back and the white belly and underwings give a flashing or twinkling effect as the birds wheel and turn. Look for white under the tail. Favors inshore waters, often appearing within sight of land. Numbers appear to be increasing in the past two decades. **Length:** 14 in. (36 cm). **Wingspan:** 33 in. (84 cm).

**AUDUBON'S
SHEARWATER**

Dark leading
edge of wing
lining

Dark
under
tail

Relatively
long tail

**Greater
Shearwater**
Shown for
comparison

The more common
Greater Shearwater
is *much* larger than
either the Manx or
Audubon's Shearwater

**MANX
SHEARWATER**

Light
leading
edge of
wing
lining

Relatively
short tail

White
under
tail

SMALL
SHEARWATERS

199

STORM-PETRELS

These "storm" birds are drawn to severe storms at sea, where they feed in the roiled waters. The name "petrel" derives from the biblical story of St. Peter walking on the water; when these small birds feed on plankton, they often dangle their feet, giving the appearance of walking on water.

WILSON'S STORM-PETREL   *Oceanites oceanicus*

Small and black with a white rump patch. Has the skimming flight of a swallow. Patters about on water's surface when feeding. If viewing up close, try to note square end of tail. At close range, feet can be seen extending beyond tail tip. Yellow toe webbing.

Perhaps the world's most common bird. Roams the high seas worldwide, feeding on planktonic life. Can be chummed near boats by tossing out fish oil and puffed rice. Breeds in burrows along Antarctic Coast. Common in North Atlantic waters in summer. **Length:** 7.5 in. (19 cm). **Wingspan:** 16 in. (41 cm).

LEACH'S STORM-PETREL   *Oceanodroma leucorhoa*

Very similar to the Wilson's Storm-Petrel, but note forked tail. Dark chocolate brown with grayish shoulders. Black toe webbing. Unique bouncing, veering flight pattern; from a distance, Leach's Storm-Petrels suggest black butterflies fluttering erratically at the water's surface. Uncommon.

Breeds along North Atlantic Coast in burrows dug into dirt banks and in rock crevices along shorelines and offshore islands. Leaves and returns to burrow under cover of night. Feeds far out to sea. Winters south of breeding areas, with scattered records of lone birds seen at sea as far north as New Jersey. **Length:** 7.5 in. (19 cm). **Wingspan:** 19 in. (48 cm).

WILSON'S STORM-PETREL

White upper tail

Rounded or squared tail

LEACH'S STORM-PETREL

STORM-PETRELS

Dark line divides white rump patch

Notched tail feathers

Pale upper wing coverts

White barely visible on rump or sides

201

RARE STORM-PETRELS

Two rare species of storm-petrel very similar in shape to the common Wilson's and Leach's Storm-Petrels described on the preceding pages. Look for these unusual petrels mixed in with masses of the more common petrels, Northern Fulmars, and shearwaters.

WHITE-FACED STORM-PETREL   *Pelagodroma marina*

A rare, deep ocean petrel. Grayish with white underparts, gray rump, and white face with strong white eye line. Flies low to the water with feet dangling, bouncing up and down like a puppet on a string. While feeding, it often swings side to side like a pendulum. Mainly a bird of southern oceans. Occurs in our area principally from late July through September, far offshore, mixed in with other petrels at upwellings along the continental shelf. **Length:** 7.5 in. (19 cm). **Wingspan:** 18 in. (46 cm).

BAND-RUMPED STORM-PETREL   *Oceanodroma castro*

Very similar in appearance to Wilson's Storm-Petrel but larger and generally darker. Area of white on rump is more extensive. Often overlooked because it mixes in with masses of Wilson's. Flight is shearwater-like, with deep wingbeats followed by long glides on stiff wings. Search any large gatherings of petrels far offshore for this uncommon species. Known to occur as far north as the Grand Banks, where it is rare or casual, and in the Gulf Stream of the southern Atlantic Coast. **Length:** 9 in. (23 cm). **Wingspan:** 20 in. (51 cm).

WILSON'S STORM-PETREL

Shown for comparison

WHITE-FACED STORM-PETREL

The only storm-petrel with extensive white below

Smaller and darker than the White-faced, with slimmer wings

White face and throat

Legs extend well beyond tail

Broad, short wings, unlike the slimmer wings of Wilson's

Divided rump patch; lighter carpal bars on wings

Broader, shorter wings than Leach's

Tail squared, not notched as in Leach's

LEACH'S STORM-PETREL

Shown for comparison

BAND-RUMPED STORM-PETREL

Carpal band on wings is darker than in Leach's

BROWN PELICAN

Pelecanus occidentalis

Description: This familiar species is one of our largest birds, larger even than the Northern Gannet. The pouched bill is unique to pelicans. Adult is brownish gray with fine gray feathering. Rich chestnut hind neck, yellow crown and throat. Juvenile is mottled brown on back and light brown on breast and belly. Immature bird has a mix of juvenile and adult characteristics; three years are required to develop full adult plumage. **Length:** 48 in. (1.2 m). **Wingspan:** 79 in. (2.1 m).

Habits: Found around marinas perched on posts, boats, and docks as well as on breakwaters and offshore rocks. Often seen plunge-diving for food or skimming water in undulating lines of birds following wave troughs. At long distances the splashes from these dives are a good field mark for feeding pelicans. Pelicans will follow boats (especially fishing boats) a moderate distance offshore looking for food, but they are not true oceanic birds. Brown Pelicans are very common on the east and west coasts of Florida and are common along the southern Atlantic seaboard north to North Carolina. The population of Brown Pelicans has been expanding northward, and the birds are now seen regularly as far north as New Jersey.

Juvenile plumage
(about six months old)

Adult nonbreeding
(in transition from
nonbreeding to
breeding plumage)

P. LYNCH

BROWN PELICAN

Adult breeding
Summer plumage

Juvenile
First year

Adult nonbreeding
Winter plumage

Adult breeding
Summer plumage

205

WHITE PELICAN

Pelecanus erythrorhynchos

Description: This huge bird is white at all ages with black trailing wing edges and a rich yellow bill. Often soars very high, and in large flocks. Breeding adults often have a caruncle, known as a "centerboard," along the top of the bill, which can remain in place well into the fall. **Length:** 60 in. (1.5 m). **Wingspan:** 108 in. (2.7 m).

Habits: Unlike the Brown Pelican, White Pelicans do not feed by plunge-diving. They float on the water and scoop food up with their huge bills. Groups hunt fish in unison, forming a united front and driving fish into shallow water where they can be caught more easily. The White Pelican's breeding range is gradually extending eastward to include sites along the Texas Gulf Coast. It is a regular winter species in southern Florida and is expanding its range east and north along the Atlantic Coast.

Similar species: Two other coastal soaring species might be mistaken for White Pelicans. White Ibis are white with black wingtips but are *tiny* compared to pelicans. Wood Storks often soar in large numbers, but they have long legs trailing behind, and the neck and head are thin and held straight out from the body. In Texas, the rare Whooping Crane is all white with black wing tips, whereas the entire read edge of the White Pelican's wing is black.

The White Ibis, White Pelican, and Wood Stork are all black-and-white soaring birds with similar silhouettes in flight. The flight profiles below emphasize the distinct *shapes* and coloration of the birds *but are not drawn to scale*: White Pelicans have almost twice the wingspread of Wood Storks, and White Ibis are much smaller than Wood Storks.

Not to scale

WHITE PELICAN

Immature

Immature birds
have mottled backs;
mature adults have
a pure white back

**Adult
nonbreeding**

Wingspan: 108 in.

Juvenile

Adult breeding

Yellow crest

Breeding adults
often have
a caruncle
("centerboard"),
a fibrous crest
on the center
line of the bill

207

NORTHERN GANNET

Morus bassanus

Description: The largest indigenous seabird along the Atlantic Coast. Adult very large and white with black outer wings. Juvenile variable brown; young birds gradually become lighter over several years toward adulthood. Note white pointed tail and very large bill, giving the body a "double-pointed" look. At close range, the golden tint over the adult bird's head and nape may be visible. **Length:** 37 in. (94 cm). **Wingspan:** 72 in. (1.8 m).

Habits: Gannets breed in large colonies on seaward cliffs around the mouth of the Gulf of St. Lawrence, such as Bonaventure Island, Quebec, and at numerous sites along the eastern Newfoundland coast. In winter they range widely at sea, wandering south as winter approaches. When feeding they make spectacular dives into the sea from heights of 40 ft. (12 m) or more. At long distances, the splashes from these dives are a good field mark for Gannets. Gregarious, they are often seen in the company of other seabirds over shoals of smaller schooling fish. They sometimes roost on the water in loose rafts of a few dozen individuals. In winter Gannets range from about one mile offshore out to the edge of the continental shelf. However, they are often seen from shore on Cape Hatteras and other prominent coastal headlands and sea islands.

Similar species: Much larger than any gull. No gull has a long, pointed tail, and gulls never plunge headlong into the water.

First-year juvenile

Second- or third-year immature

Adult plumage (at least five years old)

Outer wings are black to the wrist, unlike the more limited black wingtips of gulls

NORTHERN GANNET

First-year immature

Adult plumage

The heavy bill, shielded nostrils, and thick head plumage are adaptations to diving from great heights into the sea

Adult

Juvenile
First year

209

MASKED BOOBY, BROWN BOOBY

Relatives of the Northern Gannet, these two booby species are tropical birds that occasionally stray north on the Gulf Stream as far as the North Carolina coast. The much rarer Red-Footed Booby (see pages 212–213) is even more firmly tropical and is most often seen off the southern tip of Florida.

MASKED BOOBY
Sula dactylatra

Closely resembles the Northern Gannet but has black along the entire trailing edge of the wing rather than just on the tips. Bill is bright yellow, not grayish. Juveniles have all-white underwings separating them from adult Brown Boobies and have a brown head with a white neck collar. Much rarer than the Brown Booby. **Length:** 32 in. (81 cm). **Wingspan:** 64 in. (1.6 m).

BROWN BOOBY
Sula leucogaster

The most likely of the three boobies to be seen with regularity off the Atlantic Coast. Smaller than the Northern Gannet, with a longer tail. Chocolate brown back. White underparts, with sharp cutoff at chest between underparts and all-brown head. Bill bright yellow. Juveniles are a uniform cocoa brown all over with lighter brown underbelly. Check all trawlers coming in off the open ocean—this species will often cover the rigging or follow in the ship's wake. **Length:** 30 in. (76 cm). **Wingspan:** 60 in. (1.5 m).

BROWN BOOBY
Adult
Juvenile
First year

Adult

MASKED BOOBY

Juvenile
First year

MASKED BOOBY

Adult

Black
secondaries
form a dark
trailing edge

Black tail

Juvenile
First year

**NORTHERN
GANNET**
Shown for
comparison

White tail

Northern Gannets
are much larger than
boobies but appear
in the same areas in
winter

White
secondaries

BROWN BOOBY

Juvenile
First year

Adult

Brown
Boobies are
uniformly
dark above

BOOBIES

RED-FOOTED BOOBY

Sula sula

The smallest and rarest of the boobies has two color phases. Dark adults are uniformly light brown with an all-white tail. The white phase is all white with dark wingtips and black running along the trailing edge of the wing. Red feet are of no help in making an identification unless the bird is seen in flight or while sitting on a buoy or other floating object. Juveniles are grayish brown overall. **Range:** Rare scattered sightings along the East and Gulf Coasts, but the best opportunity to see a Red-Footed Booby is in the offshore waters of extreme southern Florida and the Florida Keys. **Length:** 28 in. (71 cm). **Wingspan:** 60 in. (1.5 m).

Similar species: To separate the white boobies and the mature Northern Gannet, concentrate on the tail color and the trailing edge of the wing. In boobies the tail will help you separate the Masked and Red-Footed Boobies. The white trailing edge of the wing is the quickest way to separate adult gannets from the white boobies.

Dark
Adult

White
Adult

RED-FOOTED BOOBY

One bird species but two distinct varieties. The Red-Footed Booby occurs in two forms, a dark adult form ("dark morph") and a white adult form ("white morph"). Juvenile Red-Footed Boobies are a uniform drab brown, resembling a slightly darker form of the dark adult.

RED-FOOTED BOOBY

White Adult

Dark Adult

White tail

Red feet

Black
trailing
edge

White tail

Red feet

Black
trailing
edge

**NORTHERN
GANNET**
Adult

ws: 72 in.

**MASKED
BOOBY**
Adult

ws: 64 in.

**RED-FOOTED
BOOBY**
White adult

ws: 60 in.

White tail

White
trailing
edge

Black tail

Black
trailing
edge

White tail

Red feet

Black
trailing
edge

CORMORANTS

Cormorants are large, dark waterbirds that are increasingly common along the Atlantic Coast. Not truly oceanic, these birds are familiar inhabitants of most harbors from Canada to Florida and along the Gulf Coast. Unlike other marine birds, cormorants dry their wings by standing with wings outstretched. In flight cormorants can look like geese, but their dark coloration and "uphill" angle of the body in flight differ from geese and Brant.

GREAT CORMORANT
Phalacrocorax carbo

Bulkier body and proportionately shorter and thicker neck than the more common Double-crested Cormorant. In summer, flanks show large white patches. Immature and some winter birds have white belly and dark throat and neck. This is a northern cormorant. In winter, wanders south in small numbers along the Atlantic Coast to northern Florida.
Length: 36 in. (91 cm). **Wingspan:** 63 in. (1.6 m).

DOUBLE-CRESTED CORMORANT
Phalacrocorax auritus

The common cormorant of the Atlantic and Gulf Coasts. Usually encountered sitting on breakwaters outside harbors. Erect stance, long, angular neck, and uptilted bill are all characteristic. An inshore species that rarely ventures out of sight of land.
Length: 32 in. (81 cm). **Wingspan:** 52 in. (1.3 m).

Double-Crested Cormorant, Anhinga Trail, Everglades National Park, Florida.

P. LYNCH

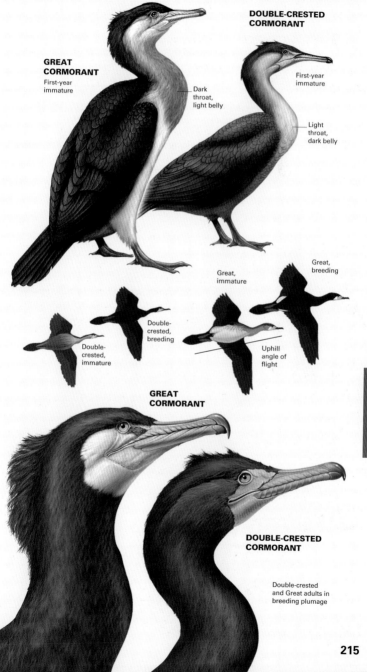

GREAT CORMORANT
First-year immature

Dark throat, light belly

DOUBLE-CRESTED CORMORANT
First-year immature

Light throat, dark belly

Double-crested, immature

Double-crested, breeding

Great, immature

Great, breeding

Uphill angle of flight

GREAT CORMORANT

DOUBLE-CRESTED CORMORANT

Double-crested and Great adults in breeding plumage

ANHINGA

Anhinga anhinga

Description: Common inhabitant of coastal and inland freshwater areas, a familiar species in the South. The long, thin neck is often all one sees above the water as it hunts for food, giving the bird its local name of "snake bird." Bill long and sharply pointed. Adult male head and neck are black, and the back is beautifully patterned with silver markings. Females and immatures are a rich golden brown above, with darker brown belly and black-and-white back.
Length: 35 in. (89 cm). **Wingspan:** 45 in. (1.4 m).

Habits: Often seen soaring over wetlands. The long, fanned tail looks like a turkey's tail and gives the Anhinga another local name, "water turkey." In flight, often beats its wings rapidly several times before continuing to soar. Commonly seen spreading its wings to dry and stretching out its long neck on an exposed branch or other perch near water. Common in the southern part of its range and appears to be edging its way slowly north along the East Coast.

Similar species: The Double-Crested Cormorant has a much shorter, stockier neck, a thicker bill that is down-turned at the tip, and a heavier body than the sleek, elongate Anhinga. Both species often occur in the same habitat.

Anhinga

Double-Crested Cormorant

Anhingas and Double-Crested Cormorants are often seen together in southeastern coastal and wetland habitats. Note the much stouter body plan and proportionately shorter neck of the cormorant and the slender, "snakelike" appearance of the Anhinga's head and neck.

ANHINGA

Adult male

Adult female

Adult male

Adult male

Wingspan: 45 in.

Females and immature birds are very similar, but immatures are more brown overall

Adult female

Adult breeding male

The classic Anhinga spread-winged pose may function both to dry the wings and as a breeding display

217

BLACK-CROWNED NIGHT-HERON
Nycticorax nycticorax

Bulky, hunched appearance. Often emits a loud "wock" call when flying to a feeding area at dusk or night. Normally feeds on small crabs and invertebrates. Sometimes preys on seabird nesting colonies on offshore islands. Active at dawn and dusk. Watch for this bird tucked in among breakwater rocks as your boat leaves the harbor. The most widespread heron in the United States. **Length:** 25 in. (64 cm). **Wingspan:** 44 in. (1.1 m).

YELLOW-CROWNED NIGHT-HERON
Nyctanassa violacea

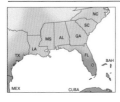

Sleeker than the Black-Crowned Night-Heron. Adult slate gray overall with fine silver back markings. Large white "slash" behind eye. Crown varies from dull yellow to chrome yellow. In flight, look for longer legs extending well beyond tail tip. Immature night-herons are more difficult to separate. The Yellow-Crowned has a heavier, all-gray bill; the Black-Crowned has a trimmer bill with a yellow lower mandible. The Yellow-Crowned is much more common than the Black-Crowned the farther south one goes from the Northeast. Found along entire Atlantic and Gulf Coasts, as well as inland in wetlands. **Length:** 24 in. (61 cm). **Wingspan:** 44 in. (1.1 m).

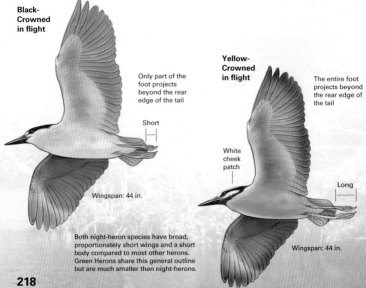

Black-Crowned in flight

Only part of the foot projects beyond the rear edge of the tail

Short

Wingspan: 44 in.

Yellow-Crowned in flight

The entire foot projects beyond the rear edge of the tail

White cheek patch

Long

Wingspan: 44 in.

Both night-heron species have broad, proportionately short wings and a short body compared to most other herons. Green Herons share this general outline but are much smaller than night-herons.

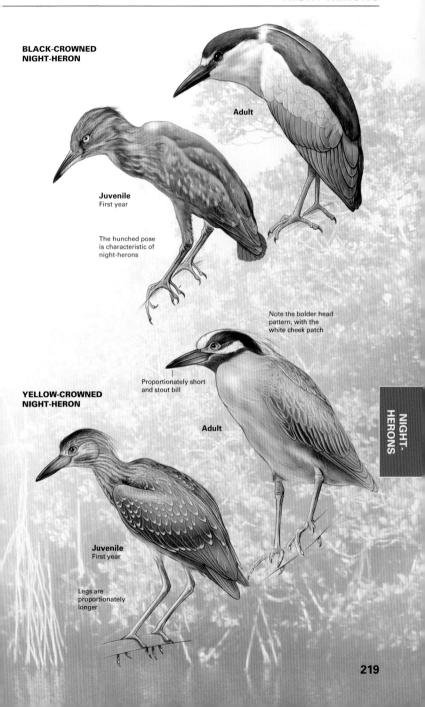

BLACK-CROWNED NIGHT-HERON

Adult

Juvenile
First year

The hunched pose is characteristic of night-herons

Note the bolder head pattern, with the white cheek patch

Proportionately short and stout bill

YELLOW-CROWNED NIGHT-HERON

Adult

Juvenile
First year

Legs are proportionately longer

NIGHT-HERONS

219

GREEN HERON, TRICOLORED HERON

GREEN HERON
Butorides virescens

Smallest and most common of the herons. Adult has a distinct greenish cap and back with a rusty neck and collar area that sets off the cap. Immatures are brown with white streaking on chest. Because of its small size, the only species that could be confused with the immature Green Heron is the Least Bittern, but in all color phases the Least Bittern has an all white-underbelly and bold rufous cream wings. Widespread in all US wetlands and coasts, Bahamas. **Length:** 18 in. (46 cm). **Wingspan:** 26 in. (66 cm).

TRICOLORED HERON
Egretta tricolor

As the name implies, the Tricolored Heron displays a mix of three distinct colors: gray, brown, and white. In all plumages the underbelly is white. Bill very long and thin. In breeding plumage, skin around bill base and eye is deep blue. Juveniles show the same basic pattern but are much paler and more rufous overall. Found along eastern seaboard from New England to Florida and Gulf Coast. **Length:** 26 in. (66 cm). **Wingspan:** 38 in. (97 cm).

Green Heron in White Mangrove. Merritt Island National Wildlife Refuge, Florida. Look for Green Herons perched in low foliage along canal banks, mangrove channels, and over water.

P. LYNCH

GREEN HERON

Adult

Immature
First year

Adult

Adult

TRICOLORED HERON

Juvenile
First six months

Dark breast and head
contrasts with light
belly and sides

Adult

HERONS

LITTLE BLUE HERON, REDDISH EGRET

LITTLE BLUE HERON
Egretta caerulea

Small heron the size of a Snowy Egret. Adult bluish gray overall with maroon tones to head, neck, and shoulders. In their first year, birds have two color patterns: up to six months they are all white and are most often confused with Snowy Egrets. Juvenile Little Blue is all white, has greenish legs and the bill and lores are gray. The stage between juvenile and adult plumage is called "piebald," white with splotches of gray in wings and on back. Found throughout the eastern United States, most common in the South. **Length:** 24 in. (61 cm). **Wingspan:** 40 in. (102 cm).

REDDISH EGRET
Egretta rufescens

Two color phases, or morphs. Dark morph blue-gray with rich reddish head and neck. Bill heavy and bicolored, pink with black tip. Often keeps its head crest raised. Lead gray legs. Juveniles are a pale rendition of adult plumage. White morph all white but with bicolored pink and black-tipped bill and gray legs. Juveniles have dark gray legs, a heavy, dark bill, and gray lores. A true southern heron, concentrated in southern Florida and along Gulf Coast. **Length:** 30 in. (76 cm). **Wingspan:** 48 in. (1.2 m).

Comparison of the Snowy Egret with the very similar immature plumage of the Little Blue Heron. Leg color is often the easiest distinction to make: black in the Snowy Egret, green in the immature Little Blue Heron. Also note the contrast between the Snowy Egret's bright yellow lores and black bill and the drab grayish lores and bill of the Little Blue Heron.

P. LYNCH

LITTLE BLUE HERON

Gray bill and lores

Black bill, yellow lores

Snowy Egret Adult
For comparison

Bicolored bill

Reddish Egret, White Morph
For comparison

Juvenile
First six months

Adult

All-dark belly is unique among small herons

"Piebald" Immature Little Blue Heron
Six to twelve months old

Adult breeding

White Morph
Note the bicolored bill

REDDISH EGRET

Juvenile
First six months

Pale, dull-colored version of the adult

HERONS, EGRETS

223

WHITE EGRETS

CATTLE EGRET
Bubulcus ibis

Small, white, with short, dark legs. Orange nape, bill, and legs in breeding season. Common to abundant in coastal grassy areas, fields, and wetlands and along roadsides. Roosts in large colonies. The Cattle Egret has spread throughout the United States but remains most abundant in the South. **Length:** 21 in. (53 cm). **Wingspan:** 36 in. (91 cm).

SNOWY EGRET
Egretta thula

Small, white overall, with black legs and bright yellow feet. Bill long, thin, and all black. Lores yellow to yellow-green in breeding season. The immature has greenish yellow legs and feet and a thin black bill. Could be confused with the juvenile Little Blue Heron, which is also all white but has a heavier, gray bill and greenish gray legs. Found throughout the East and Gulf Coasts. **Length:** 24 in. (61 cm). **Wingspan:** 42 in. (107 cm).

GREAT EGRET
Ardea alba

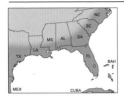

The largest egret. All white with heavy yellow bill and black legs. Lores deep green in breeding season. Common in shoreline marshes, on offshore islands, and in shallows of beaches. Seen in large concentrations when coastal fish move into marshlands. Found throughout the eastern and Gulf seaboards, increasing in numbers each year as its range expands northward. **Length:** 40 in. (102 cm). **Wingspan:** 51 in. (1.3 m).

Cattle Egrets in flight near Lake Mattamuskeet National Wildlife Refuge, Hyde County, North Carolina. Note the proportionately short neck and legs and the orange nape and bill.

P. LYNCH

CATTLE EGRET

Juvenile

Adult nonbreeding

Adult breeding

SNOWY EGRET

Juvenile

Adult breeding

Adult nonbreeding
Immature has similar plumage

GREAT EGRET

Adult breeding
Note the intense colors of the lores and bill

225

GREAT BLUE HERON   *Ardea herodias*

GREAT BLUE HERON

This common bird of shore, marshland, and inland waters is our largest heron. Immatures have streaking on throat and upper chest, lack long head plumes, and are more uniform gray all over. Found throughout the United States and well into Canada. **Length:** 46 in. (1.2 m). **Wingspan:** 72 in. (1.8 m).

INTERMEDIATE MORPH ("WÜRDEMANN'S HERON")

Found in south Florida and especially in the Florida Keys, the intermediate morph has the body conformation of the typical Great Blue Heron, but the head, neck, throat, and chest are white.

WHITE MORPH ("GREAT WHITE HERON")

This white morph of the Great Blue Heron is a little larger, with a very heavy yellow bill. The head plumes, if present, are very short. Could be confused with the much smaller Great Egret (see below). Most common in the Florida Keys but slowly extending its range northward into southern Florida.

The shape and proportions of the bill can help you to distinguish Great Egrets from the similar but much larger "Great White Heron" form of the Great Blue Heron.

GREAT EGRET
Great Egrets have relatively long yellow bills with less height at the base than that of the much larger Great White Heron

Lores are usually bright yellow but may be bright yellow-green or even blue at the height of breeding season

"GREAT WHITE HERON"
Great White Herons have massive, more triangular bills that are often dark around the nostrils and quite thick at the base

Lores predominantly gray

226

Immature

GREAT BLUE HERON

"WÜRDEMANN'S HERON"

Würdemann's is an intermediate form between the typical Great Blue Heron and the larger Great White Heron

A typical Great Blue Heron in flight

"GREAT WHITE HERON"

Great White Herons are larger and have heavier bills than the average Great Blue Heron

GREAT BLUE HERON

227

WOOD STORK, FLAMINGO

WOOD STORK
Mycteria americana

The only true stork found in the United States. Juveniles have a feathered neck and back of the head. Hunts by standing or slowly walking in the water with bill opened so any fish that swims into it is snapped up. Could be confused in soaring flight with the White Pelican. However, pelicans fly in uniformly structured lines, whereas storks form unstructured "kettles" of circling birds. Mainly concentrated in the Florida and Gulf Coast areas, but in recent years more wandering birds have been seen northward into the mid-Atlantic states and even Canada. **Length:** 40 in. (102 cm). **Wingspan:** 61 in. (1.5 m).

GREATER FLAMINGO
Phoenicopterus ruber

Though one of the best known birds, flamingos have never been a typical bird of Florida. Our flamingos are usually storm-driven wanderers from Mexico or semiwild escapees from captive flocks. A small, permanent population exists in Flamingo Bay off the southern Everglades. Young birds are much duller pink. Captive birds lose their vibrant color unless properly fed. Confined to extreme south Florida and Texas, though scattered individuals are found throughout Florida. **Length:** 46 in. (1.2 m). **Wingspan:** 60 in. (1.5 m).

Wood Stork in flight, Merritt Island National Wildlife Refuge, Florida. Note the splayed primary feathers at the wingtips and the legs that project well beyond the tail.

P. LYNCH

WOOD STORK

Immature
Young birds gradually lose the neck feathering. Adults have bare necks and heads.

Adult

GREATER FLAMINGO

Adult

Immature

Captive flamingos gradually lose their pink color unless fed a special diet that mimics their wild foods

ROSEATE SPOONBILL

Ajaia ajaja

A large bird with a unique flattened, spoon-shaped bill and a beautiful pink and white body. The adult shoulder area is a deep crimson, and in flight the tail reveals a rich orange color. Immature roseate spoonbills are pale versions of adults. The eyes are deep red, and the face of the adult is naked. Spoonbills feed by holding their bill open in the water and moving their head from side to side. The sight of flocks returning to roost in the mangroves at sunset make for an indelible memory. Their unique color could only be confused with a young flamingo, but Greater Flamingos are rare outside of extreme south Florida, and flamingos are much taller birds. Spoonbills are found in the south half of Florida and along the entire Gulf Coast, with scattered records along the East Coast north to New Jersey.

Length: 32 in. (81 cm). **Wingspan:** 50 in. (1.3 m).

Roseate Spoonbill populations were decimated in the early 1900s by plume hunters seeking their gorgeous pink feathers. Spoonbills have reclaimed much of their former range but are still vulnerable to habitat destruction and wetland development along the Florida and Texas coasts.

N. PROCTOR

ROSEATE SPOONBILL

Adult

Juvenile

Juvenile

It takes three to four
years for spoonbills to
assume the intense pink
and scarlet plumage of
the mature adult

**Second
year**

Adult

ROSEATE
SPOONBILL

231

IBISES

WHITE-FACED IBIS
Plegadis chihi

A long-legged bird much like a heron, but with a long, downcurved bill. Note the bright crimson eye and, in breeding plumage, a deep red face bordered completely by a white outline that runs behind the eye. This western bird is extending its range east along the Gulf Coast. **Length:** 24 in. (61 cm). **Wingspan:** 37 in. (94 cm).

GLOSSY IBIS
Plegadis falcinellus

Adult rich copper overall with bright reflective green in wings and top of head. The White-faced Ibis is nearly identical but is always duller in overall color, is slightly larger, and always has the deep crimson-colored eye. Found along East and Gulf Coasts. **Length:** 23 in. (58 cm). **Wingspan:** 36 in. (91 cm).

WHITE IBIS
Eudocimus albus

Adult pure white with red bill and legs. Common to abundant resident of wetlands, marshes, and offshore islands of southeastern and Gulf Coasts. White Ibises often forage along roadsides, on medians, and even in parks and campgrounds. **Length:** 25 in. (64 cm). **Wingspan:** 38 in. (97 cm).

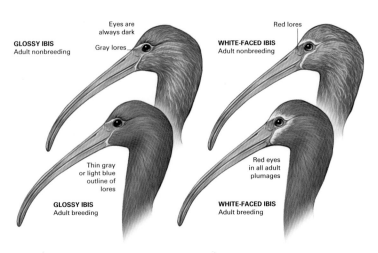

GLOSSY IBIS
Adult nonbreeding

Eyes are always dark

Gray lores

WHITE-FACED IBIS
Adult nonbreeding

Red lores

Thin gray or light blue outline of lores

GLOSSY IBIS
Adult breeding

Red eyes in all adult plumages

WHITE-FACED IBIS
Adult breeding

Adult nonbreeding

WHITE-FACED IBIS

Adult breeding

Adult nonbreeding

GLOSSY IBIS

Adult breeding

WHITE IBIS
Juvenile

SCARLET IBIS

Eudocimus ruber

Introduced from South America and closely related to the White Ibis. Very rare outside southern Florida, and unusual and scattered there except in parks and zoos.

WHITE IBIS
Adult breeding

Dark primary feather tips

SWANS

TUNDRA SWAN
Cygnus columbianus

Native swan, all white, with a black bill that shows a yellow "teardrop" mark near the eye. Neck is usually held straight up instead of in an S-curve, as in the Mute Swan. Juvenile is grayish brown with a pink bill with a bright pink edge. Breeds in the far north and migrates to the East and West Coasts. East Coast birds winter on the Virginia and North Carolina coasts. **Length:** 60 in. (1.5 m). **Wingspan:** 65 in. (1.7 m).

MUTE SWAN
Cygnus olor

Instantly recognizable by its huge size, pure white color, and habit of holding its neck in a graceful S-curve. Bill is deep orange with a black base. Male bill has a larger swollen knob at the base. Juveniles can be either gray or white with a pale black bill. Introduced to North America by a New York opera lover enamored of the swans in Wagnerian operas. Now found on the Atlantic Coast from New England to North Carolina and expanding southward. **Length:** 60 in. (1.5 m). **Wingspan:** 75 in. (1.9 m).

Mute Swans always impress watchers because of their graceful carriage. But remember: do not feed bread to any waterfowl. Bread can cause developmental problems such as "angel wings," a twisting distortion of the wing bones and feathers that can eventually prevent the birds from flying. Do not allow small children to feed swans by hand. These large birds can be very aggressive, especially in breeding season or when protecting their young.

P. LYNCH

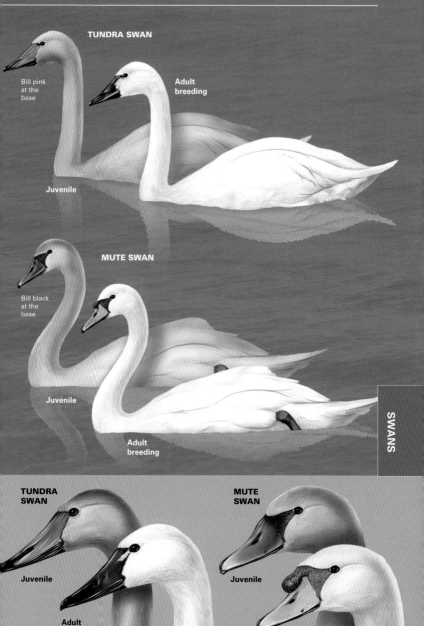

TUNDRA SWAN

Bill pink at the base

Adult breeding

Juvenile

MUTE SWAN

Bill black at the base

Juvenile

Adult breeding

SWANS

TUNDRA SWAN

Juvenile

Adult breeding

MUTE SWAN

Juvenile

Adult breeding

235

COASTAL GEESE

CANADA GOOSE
Branta canadensis

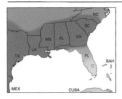

Common goose of coastal waters, inland ponds, bays, and marshes. Widely used as an "ornamental" goose for ponds and other freshwater habitats. These introduced birds are difficult to separate from the true wild and migratory Canada Geese. Introduced birds are more often than not the large subspecies called the *Common Canada Goose* (length typically 45 in., 111 cm), whereas wild Canada Geese, which migrate south in fall, spend the winter there, and then head back north to nest in spring are usually *Lesser Canada Geese* (length typically 36 in., 91 cm). Found throughout United States and Canada. On East Coast, range extends to northern Florida and expands south each year. Also found all along Gulf Coast from Florida Panhandle to Mexican border. **Length:** 36–45 in. (91–111 cm). **Wingspan:** To 60 in. (1.5 m).

BRANT
Branta bernicla

Small coastal and offshore goose resembling a small Canada Goose. However, back is darker brown to almost black, and black neck color extends down to a sharp cutoff mark on lower breast. Instead of the Canada Goose's bold chinstrap, the Brant has a small necklace of white lined with black. Bill is small and black. Breeds in the far north above Hudson Bay and migrates to coastal waters for fall and winter. Principal winter grounds range from the New England coast south to the northernmost coast of Georgia. There are scattered inland records and increased sightings farther south along the East Coast and on the Gulf Coast. **Length:** 26 in. (66 cm). **Wingspan:** 42 in. (107 cm).

Note: In the 1970s and 1980s the Brant population was in serious decline because its favorite food, Eelgrass (*Zostera*), was decimated by a fungus. Fortunately, the birds altered their feeding habitats and now feed on the marine algae Sea Lettuce (*Ulva lactuca*) and the fresh shoots of *Spartina* salt grass, and the Brant population of the East Coast has slowly begun to recover.

**CANADA
GOOSE**

*Sexes are alike
in all plumages*

BRANT

*Sexes are alike
in all plumages*

COASTAL GEESE

GREATER WHITE-FRONTED GOOSE
Anser albifrons

Medium-sized (smaller than most Canada Geese), with plumage of fairly uniform tans and browns, bright white forehead, and pink bill. Bright orange legs and feet are often the most noticeable field mark. Migrates primarily across central North American flyway and winters on Texas and Louisiana coasts. Each of the four major geese species has several distinct geographic races, so you'll often see some variations in plumage patterns and body size in large flocks. The geese species also regularly hybridize, producing intermediate forms, mostly by mixing with Canada Geese. In our area these geese are seen primarily on Texas and Louisiana coasts in winter, but a few regularly reach the East Coast and eastern Gulf Coast and Florida. **Length:** 28 in. (71 cm). **Wingspan:** 54 in. (1.4 m).

SNOW GOOSE AND "BLUE GOOSE"
Chen caerulescens

Snow Geese appear in two major forms, or morphs: the classic white Snow Goose and a dark morph once classified as a separate species, the "Blue Goose." Juveniles of both morphs are darker, duller versions of adults and have gray rather than white heads. Migrates from Canadian Arctic breeding grounds primarily along central North American flyway, and bulk of population winters on Gulf Coast west of Florida Panhandle. Smaller numbers winter along East Coast from Delmarva Peninsula south to Georgia. Unusual on either Florida Coast, except in westernmost panhandle area. **Length:** 29 in. (74 cm). **Wingspan:** 55 in. (1.4 m).

GREATER WHITE-FRONTED GOOSE

White-Fronted Geese, Snow Geese, and Canada Geese sometimes interbreed and produce hybrids. Shown is a White-Fronted x Canada Goose hybrid.

SNOW GOOSE
White morph of the Snow Goose

"BLUE GOOSE"
Dark morph of the Snow Goose

SCOTERS, COMMON EIDER

Scoters are dark sea ducks that spend their winters in coastal waters, sometimes many miles offshore. They dive to feed on mollusks and crustaceans. Their flight is strong, direct, and close to the waves. Eiders are large, primarily northern sea ducks that enter our area only around Virginia and Cape Hatteras in the winter months.

BLACK SCOTER
Melanitta nigra

Male all black with bright orange-yellow knob on bill. Female brown with distinct pale cheeks, giving capped appearance. Forms small groups in winter along coastlines, but never in large rafts as formed by other scoter species. **Length:** 15 in. (38 cm). **Wingspan:** 35 in. (89 cm).

WHITE-WINGED SCOTER
Melanitta fusca

Male all black with white wing patch and white teardrop mark behind eye. Female brown with white wing patch and two dusky head patches at base of bill and on cheek. May form very large rafts off islands and coastlines. Perhaps the most common of the three scoters. **Length:** 17 in. (43 cm). **Wingspan:** 39 in. (100 cm).

SURF SCOTER
Melanitta perspicillata

Male black with distinct white patches on forehead and nape. Brightly colored bill. Female brown with dusky white splotches behind bill and on side of head, dusky patch on nape. Occurs in small to large rafts. Often swims with tail sticking up. **Length:** 15 in. (38 cm). **Wingspan:** 34 in. (86 cm).

COMMON EIDER
Somateria mollissima

A very large sea duck. Male has white back and chest and black underparts. Female light brown. Young male brown with white chest. Up close, note sloping forehead, black cap, and very long bill processes running up toward eye. Can occur in tremendous numbers. Flocks are often seen flying low over the water in long, undulating skeins. **Length:** 24 in. (61 cm). **Wingspan:** 43 in. (1.1 m).

BLACK SCOTER

Female

Male

WHITE-WINGED SCOTER

Female

Male

SURF SCOTER

Female

Male

Female

Male

COMMON EIDER

241

BAY DUCKS

Bay ducks winter in sheltered bays and harbors along the coast or in the lee of offshore islands. Goldeneyes are less marine in their habits but may be found in shallow coastal waters and estuaries.

GREATER SCAUP
Aythya marila

The most marine of the bay ducks. In flight, white wing patches extend well into outer flight feathers. Forms massive rafts of thousands up to several miles offshore. Dives to feed on shellfish in shallow water. Numbers have decreased dramatically in the past two decades. **Length:** 18 in. (46 cm). **Wingspan:** 32 in. (81 cm).

LESSER SCAUP
Aythya affinis

Very similar to the Greater Scaup in all plumages but smaller, with more peaked head silhouette and smaller bill. It is often easiest to separate the two species by habitat: the Lesser favors inner harbors near shores and rarely moves far offshore. Any scaup more than a mile offshore is almost certainly a Greater. **Length:** 16 in. (41 cm). **Wingspan:** 29 in. (74 cm).

COMMON GOLDENEYE
Bucephala clangula

Male white with dark back and green head and white circle between eye and bill. Female gray with chocolate brown head. Winters in sheltered coastal areas, sometimes in large flocks. Stays well inshore in bays and harbors. Winters south to Gulf Coast; not found in most of Florida. **Length:** 18 in. (46 cm). **Wingspan:** 32 in. (81 cm).

GREATER SCAUP

Female

Male

LESSER SCAUP

Female

Male

COMMON GOLDENEYE

Female

Male

LONG-TAILED DUCK

Clangula hyemalis

In winter plumage, white with dark chest. Male has exceptionally long tail plumes. In flight rocks from side to side, giving appearance of white (back) then black (chest). One of the fastest ducks in flight. Favors inshore waters and can form large, very noisy flocks. Huge rafts of this duck sometimes occur offshore in winter. **Length:** 20 in. (51 cm). **Wingspan:** 28 in. (71 cm).

BUFFLEHEAD

Bucephala albeola

A very small diving duck with a rounded head and body silhouette. Male shows strong black-and-white pattern with black back, white underparts, and large white patch on head. Female a duller gray echo of male. Prefers sheltered bays and harbors and is usually seen in small mixed groups of four to eight males and females, not in large flocks. Also found inland on lakes, ponds, and rivers. **Length:** 13 in. (33 cm). **Wingspan:** 21 in. (53 cm).

RED-BREASTED MERGANSER

Mergus serrator

A fish-eating diving duck with a long, serrated red bill. Both sexes have a ragged crest at nape. Flight pattern is strong and direct, usually low over the waves. More marine in its habits than the similar Common Merganser, which prefers fresh water. Favors sheltered bays, estuaries, and harbors all along East and Gulf Coasts. **Length:** 23 in. (58 cm). **Wingspan:** 30 in. (76 cm).

LONG-TAILED DUCK

Winter male has a white crown and neck

Dark cheek spot

Female

In winter, both sexes have light heads and bodies with uniformly dark wings

Male

Male

Noticeably smaller than other ducks

BUFFLEHEAD

Female

Relatively large head; strong black-and-white patterns

Male

Male

RED-BREASTED MERGANSER

Both sexes have a ragged crest

Female

Male

Slender body profile; typically flies close to the surface, in pairs or small groups

BAY DUCKS

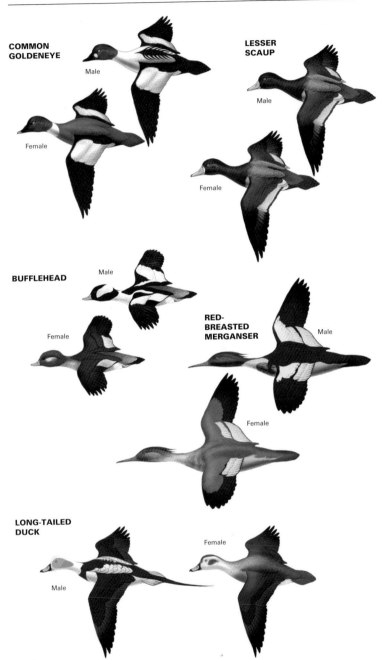

COMMON GOLDENEYE

Male

Female

LESSER SCAUP

Male

Female

BUFFLEHEAD

Male

Female

RED-BREASTED MERGANSER

Male

Female

LONG-TAILED DUCK

Male

Female

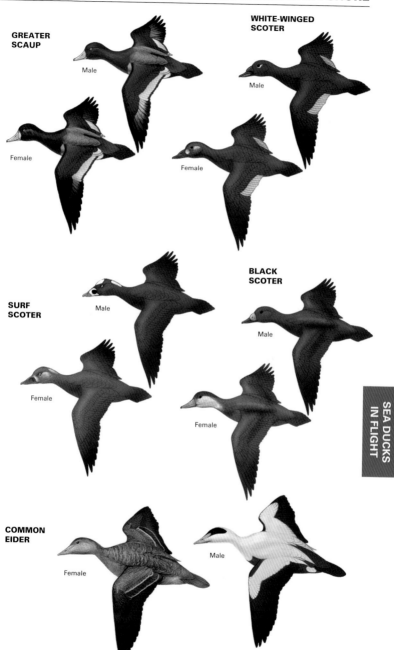

GREATER SCAUP

Male

Female

WHITE-WINGED SCOTER

Male

Female

SURF SCOTER

Male

Female

BLACK SCOTER

Male

Female

COMMON EIDER

Female

Male

OSPREY

Pandion haliaetus

An eagle-sized "fish hawk," brown-backed and white underneath. Flight silhouette with distinct "crooked" angles at wrist joints marked with dark patches. White head with brown stripe running through the eye to the shoulder. Juveniles have red-orange eyes, and their dark back feathers have distinct white edges that gradually wear off during their first year. Adults have bright yellow eyes. Ranges throughout North America and Caribbean. **Length:** 24 in. (61 cm). **Wingspan:** 64 in. (1.6 m).

Habits: Most commonly seen along coasts, marshes, and lake edges. Hovers in flight as it peers down hunting for fish, then tucks in its wings and plunges into the water feetfirst to secure its prey, which it takes to a favored feeding site such as a dead tree limb. Ospreys usually make their huge stick nests in trees, but they have shown remarkable adaptation to man-made nesting structures. Telephone poles, power line poles, and even microwave relay towers are all used for nest locations.

Similar species: Often confused with the adult Bald Eagle based on the white head. The Osprey's white underparts, however, are a quick point of separation.

White feather edges and red-orange eyes distinguish a juvenile Osprey, just fledged from the nest and still regularly fed by its parents. Outer Banks, near Corolla, North Carolina.

P. LYNCH

The long wings with prominent "crooked" wrist joints help separate the Osprey at a distance from gulls and other soaring birds

Adult

Splayed primary feather "fingers"

Prominent wrist joint with dark patch

Note the red-orange eye

Juvenile
First year

Very young Ospreys are flecked with white down

Adult

OSPREY

Juvenile
First year

Note the light edges on all the major back feathers

The Osprey was the bellwether for the misuse and damaging effects of pesticides. Since chemicals such as DDT have been banned from widespread use, Ospreys have made a remarkable comeback over the past 35 years.

249

BALD EAGLE

Haliaeetus leucocephalus

Description: Our largest bird of prey, easily separated from hawks by its much larger size and proportionately longer wings. Adult uniform dark brown with white head and tail. Immatures require four years to grow into adult plumage, during which they evolve from a uniform mottled dark brown (juvenile) into two immature years with heavily mottled underwing and belly plumage that is much lighter than that of the full adult. Found throughout North America but not present in Bahamas or Caribbean. **Length:** 32 in. (81 cm). **Wingspan:** 82 in. (2.1 m).

Habits: Full-time resident of southeastern and Gulf coasts. In winter the population swells with additional eagles that migrate to temperate southern shorelines to winter, particularly along Florida coasts and the Florida Keys. Feeds mainly on fish in our coastal regions, either caught, scavenged, or stolen from smaller birds like Ospreys. Look for Bald Eagles soaring over coastal wetlands or perched on dead trees at the edge of marshes and coastlines. Eagles will often perch on old dock pilings, channel markers, or other structures above or near water, as long as they are well away from human activity. It's always worth scanning a new shoreline view to look for perched Bald Eagles. The adults' white heads make them easy to spot even over long distances.

Similar species: The Osprey is the most likely bird to be confused with the Bald Eagle. Ospreys always have light underwings and bellies and a dark eyestripe. In flight the Bald Eagle's long, broad wings are proportionately thicker front to back than the slimmer wings of the Osprey. The Osprey also shows more pronounced bend angles at the "wrist," and Ospreys show dark wrist patches on the underwings at all ages and plumages.

Full adult plumage
Attained at about four years of age

Immature, second-year plumage

Adult plumage

Immature, first-year plumage

251

BROAD-WINGED HAWK
Buteo platypterus

A fairly common but shy summer resident of coastal woodlands and very visible fall migrant all along southeastern and Gulf coasts. In summer ranges throughout eastern North America except Florida, and during fall migration is found throughout region. **Length:** 15 in. (38 cm). **Wingspan:** 34 in. (86 cm).

RED-SHOULDERED HAWK
Buteo lineatus

Smaller, slimmer version of the more common Red-Tailed Hawk. Common all along southeastern and Gulf coasts, particularly in woodlands and marine forests on barrier islands. Often seen soaring above marshes and woodlands. Ranges throughout eastern North America. **Length:** 17 in. (43 cm). **Wingspan:** 40 in. (1.1 m).

RED-TAILED HAWK
Buteo jamaicensus

Our most common hawk. Highly visible year-round resident of coastal marshes, river mouths, barrier islands, and coastal woodlands. It is unusual *not* to see at least one soaring Red-Tail while birding along the coast. Found throughout North America, in winter along extreme south Texas coast and most of Caribbean. **Length:** 19 in. (48 cm). **Wingspan:** 49 in. (1.2 m).

AMERICAN KESTREL
Falco sparverius

Our smallest falcon. Common resident of shoreline marshes and coastal islands, where it is often seen perched on snags while scanning for prey. Kestrels also hover in flight while hunting above marshes, making them easy to identify from long distances. Ranges throughout North America, in winter along Texas coast. **Length:** 9 in. (23 cm). **Wingspan:** 22 in. (56 cm).

AMERICAN KESTREL

Adult
male

Adult
female

BROAD-WINGED HAWK

Rusty-red breast; wing linings more white than in the Red-shouldered

Broad white bands in the tail

Large, evenly spaced dark-light bands on tail

RED-SHOULDERED HAWK

Rustyred breast and underwings

Tail dark with thin white bands

Thin white bands on a dark tail

RED-TAILED HAWK

Dark "belly band"

Brick red tail

Brick red tail in the adult; note "belly band"

COASTAL HAWKS

VULTURES

BLACK VULTURE
Coragyps atratus

Abundant large, black soaring bird, very common in all coastal environments but rarely venturing from land unless crossing small bays or channels. Outer primaries are light gray, making wingtips look white in flight. Soars with relatively flat wings. Ranges throughout southeastern North America. **Length:** 25 in. (64 cm). **Wingspan:** 60 in. (1.5 m).

TURKEY VULTURE
Cathartes aura

Abundant large, soaring bird, larger than the Black Vulture, with all-dark wings held in a distinctive shallow V in flight. Red head, dark wings, and wing angle when soaring make it easy to separate from the Black Vulture. Ranges throughout southeastern North America. **Length:** 26 in. (66 cm). **Wingspan:** 67 in. (1.7 m).

TURKEY VULTURE Soars with wings held in a shallow V dihedral angle

BLACK VULTURE Soars with relatively flat wings; note the light wingtips

Black Vultures, Playalinda Beach, Cape Canaveral National Seashore, Florida.

P. LYNCH

BLACK VULTURE

Dark head

Light wingtips

TURKEY VULTURE

Dark wingtips

Red head

BLACK
VULTURE

TURKEY
VULTURE

VULTURES

255

SOARING BIRDS ALONG THE COAST

BROAD-WINGED HAWK
Buteo platypterus

Short banded tail and relatively rapid wingbeats. Mostly seen in fall migration, often in large groups. **Wingspan:** 34 in. (0.8 m).

RED-TAILED HAWK
Buteo jamaicensus

A very common coastal hawk, often seen soaring over marshes, coastal wetlands, and barrier islands. **Wingspan:** 49 in. (1.2 m).

RED-SHOULDERED HAWK
Buteo lineatus

A common resident of coastal woodlands, often seen soaring above forests, river banks, and marshes. **Wingspan:** 40 in. (1.1 m).

WHITE PELICAN
Pelecanus erythrorhynchos

The largest bird in our area, with a wingspan larger even than eagles and frigatebirds. Black trailing edges of the wings. **Wingspan:** 108 in. (2.7 m).

OSPREY
Pandion haliaetus

A long-winged bird of prey that soars over open water and dives feetfirst to catch fish at the water's surface. Note dark wrists. **Wingspan:** 64 in. (1.6 m).

MAGNIFICENT FRIGATEBIRD
Fregata magnificens

A small tropical seabird with a huge wingspan, made for soaring over open water. Buoyant, maneuverable flight pattern. **Wingspan:** 92 in. (2.3 m).

BROWN PELICAN
Pelecanus occidentalis

A very common coastal bird, more often see gliding above the waves parallel to shore, but also regularly soars over coastlines. **Wingspan:** 79 in. (2 m).

TURKEY VULTURE
Cathartes aura

A large, dark vulture that soars with fingerlike primaries at the end of its wings, which are held in a shallow V shape. **Wingspan:** 67 in. (1.7 m).

BLACK VULTURE
Coragyps atratus

A smaller vulture with conspicuous white patches near its wingtips. Short, angular tail and small dark head. **Wingspan:** 60 in. (1.5 m).

GREAT BLACK-BACKED GULL
Larus marinus

The largest of the gulls, common on southeastern coast but rare on Gulf Coast. Large white spots near its wingtips. **Wingspan:** 65 in. (0.8 m).

BALD EAGLE
Haliaeetus leucocephalus

Long, broad wings. Slow, heavy wingbeat, unlike rapid beats of smaller hawks. Adult's dark belly separates it from the Osprey. Young Bald Eagles have broader wings than the Osprey. **Wingspan:** 82 in. (2.1 m).

SOARING BIRDS ALONG THE COAST

Broad-Winged Hawk

Wingspan: 34 in.

Red-Tailed Hawk

49 in.

Red-Shouldered Hawk

40 in.

White Pelican

108 in.

Osprey

64 in.

Magnificent Frigatebird

92 in.

Brown Pelican

79 in.

Turkey Vulture

67 in.

Great Black-Backed Gull

65 in.

Bald Eagle

82 in.

Black Vulture

60 in.

PHALAROPES

These small shorebirds winter far out to sea, often in large flocks. Watch for fast-moving groups as approaching boats flush them from the water. Thousands are sometimes blown ashore by storms, creating "phalarope wrecks." At sea in winter phalaropes feed on small invertebrates and fish immediately below the ocean surface with a quick, nervous picking and spinning action quite unlike other seabirds.

WILSON'S PHALAROPE
Phalaropus tricolor

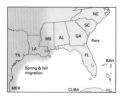

Seen in our area in migration, usually in winter (nonbreeding) plumage. Note plain medium-gray cast to back, relatively light face and head, bright white cheeks, and smudgy gray cap. If seen along shore, note the proportionately long legs, more suggestive of a Lesser Yellowlegs than other phalaropes. Feet project beyond trailing edge of tail, another useful field mark. **Length:** 9 in. (23 cm). **Wingspan:** 17 in. (43 cm).

RED-NECKED PHALAROPE
Phalaropus lobatus

Separated from the Red Phalarope by distinct dark streaking on back and overall mottled pattern of light-edged back feathers. The immature Red can show streaking but is never as contrasty as the Red-Necked. The Red-Necked bill is needlelike compared to the more substantial bill of the Red. **Length:** 8 in. (20 cm). **Wingspan:** 15 in. (38 cm).

RED PHALAROPE
Phalaropus fulicaria

A small bird with very fast, twisting flight. Separated from the Red-Necked by its plain gray back. All phalaropes have black "eyepatch" area in winter plumage. Bill much stouter than Red-Necked. Buoyant in water, like a tiny gull. **Length:** 8.5 in. (22 cm). **Wingspan:** 17 in. (43 cm).

Plain gray mantle and upper wings without a wing stripe

White rump

Long legs, with feet visible behind the tail

Very light cap with narrow eye stripe

WILSON'S PHALAROPE
Winter
(nonbreeding)

Pale gray back

Mantle and upper wings look mottled

Wing stripe

Dark rump with light sides

Dark cap with a heavy eye stripe

RED-NECKED PHALAROPE
Winter
(nonbreeding)

Back is a darker gray, with more distinct light feather edges

Wing stripe

Gray rump

Lighter cap with a heavy eye stripe

Mantle and upper wings plain gray

RED PHALAROPE
Winter
(nonbreeding)

Pale gray back

PHALAROPES

PLOVERS

Compared to other shorebirds, plovers are stockier, have shorter bills, and are more alert in their posture. They often walk rapidly, then stop and pause before continuing on.

BLACK-BELLIED PLOVER
Pluvialis squatarola

Breeding plumage boldly marked black and white. Underparts mainly black, with mottled light gray black that runs up neck to white cap. Short, black, heavy bill. Nonbreeding plumage predominantly gray, but may show areas of black on underparts. In flight, rump is white and axillaries (wing pits!) are black. Call a distinctive mournful "pee-ooo-wee." Ranges in winter along entire East and Gulf Coasts. **Length:** 11.5 in. (29 cm). **Wingspan:** 29 in. (74 cm).

AMERICAN GOLDEN-PLOVER
Pluvialis dominica

Often confused with the Black-bellied Plover, but slimmer, with more delicate bill. Breeding plumage brown flecked with gold. Cap dark brown. Underparts jet black, including under tail. White edges the black from the shoulders to the line over the eye. Nonbreeding plumage soft brown with distinct eye line and dark cap. In flight, rump brown, underwings silver-brown with no black markings. Call a mellow, high, whistled "qweedle." Breeds in high Arctic and seen in migration throughout eastern United States. **Length:** 10.5 in. (27 cm). **Wingspan:** 26 in. (66 cm).

A comparison of the American Golden-Plover (left, at St. Petersburg, Florida), and the Black-Bellied Plover (right, Padre Island National Seashore, Texas), both in nonbreeding plumage.

N. PROCTOR

BLACK-BELLIED PLOVER

Winter

White wing stripe

White rump

Breeding

Black axillaries in both winter and breeding plumages

Winter

White under tail

Breeding

AMERICAN GOLDEN-PLOVER

Dark rump

No or very faint wing stripe

Prominent white eye line

Axillaries are gray, not black

Black under tail in breeding plumage

261

SMALL PLOVERS

SNOWY PLOVER
Charadrius alexandrinus

Very pale, large-headed, with light sandy back and white underparts. Incomplete neck collar creates shoulder patches. Black forehead and black behind eye in breeding plumage. Call "koo-wee." Nests along Gulf Coast and western Florida coast. Very rare along East Coast. **Length:** 6.5 in. (17 cm). **Wingspan:** 17 in. (43 cm).

PIPING PLOVER
Charadrius melodus

Very pale. Plain face without dark cheeks. In breeding plumage, black bar across forehead and black collar around neck. These fade to pale buff in winter. Stubby black bill. Often sneaks away when approached, stopping and looking over its shoulder, relying on its sandy color for camouflage. Call "peep-low." Ranges along coast from New England south to Florida and Gulf of Mexico. **Length:** 7.5 in. (19 cm). **Wingspan:** 19 in. (48 cm). **Red List – Near Threatened**

WILSON'S PLOVER
Charadrius wilsonia

Larger than other plovers with single chest bands. Heavy black bill. Often stands erect in alert posture. Chest band black in breeding plumage, fading to brown in winter. Brown back and cheek patch. Orange smudge behind eye in breeding plumage. Dull pinkish legs. Call a sharp "keet". Found along southern coasts from Virginia through Florida and all along Gulf of Mexico. In winter, found on southern and western Florida coast and Texas coast. **Length:** 8 in. (20 cm). **Wingspan:** 19 in. (48 cm).

Piping Plover in breeding plumage, Old Lyme, Connecticut. Note the orange base of the bill.

N. PROCTOR

SNOWY PLOVER

Dark patches at the base of the neck

Breeding

Winter

Long gray legs

PIPING PLOVER

Dark collar is more complete than in the Snowy Plover

Breeding

Winter

Bright yellow legs

WILSON'S PLOVER

Heavy black bill

Breeding

Winter

Dull pinkish or yellowish legs

SEMIPALMATED PLOVER

Charadrius semipalmatus

Small, alert, with complete black collar around neck, black mask through eye, and white forehead. Bill yellow with black tip. Legs yellowish orange. Underparts white. In flight shows black-tipped tail with white sides. Call a sharp "chu-wee." Migrant through most of United States, occurring on eastern shoreline south to Florida and along Gulf Coast. In winter ranges along coast from Carolinas south. **Length:** 7.5 in. (19 cm). **Wingspan:** 19 in. (48 cm).

KILLDEER

Charadrius vociferus

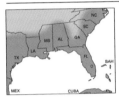

Two black neck bands, full white collar, and white at bill base and over eye. Long, tapered body, brown on back and white below. Tail rich orange, seen to best advantage when the bird flushes or uses the "broken wing act" to lure predators away from the nest site. Call a distinct "kill-dee." Prefers grassy and gravel areas and even nests on gravel roofs. One of our most common shorebirds. Ranges throughout United States, with permanent populations from New England south and along entire East and Gulf Coasts. **Length:** 10.5 in. (27 cm). **Wingspan:** 24 in. (61 cm).

The Killdeer is one of our most common and handsome shorebirds. A bird of open fields, back beaches, dunes, and any exposed open ground, including sports fields and even parking lots and flat roofs of buildings. Photographed at Corpus Christi, Texas.

N. PROCTOR

SEMIPALMATED PLOVER

Heavy complete breast band

Breeding

Long wings project beyond the tail

Winter

Brick red rump

Typical adult coloration

KILLDEER

Double breast bands

"Rufous" adult more common in the South

Brick red ("rufous") wing coverts are seen in some birds

PLOVERS

STILT, AVOCET, OYSTERCATCHER

BLACK-NECKED STILT
Himantopus mexicanus

A beautiful shorebird with striking black-and-white markings and very long pink legs. Juveniles brown and white. Bill long and very thin. Call a sharp, repeated "kip-kip-kip." Breeding range has expanded from west to include Gulf of Mexico, Florida, and southeast coast north to Virginia. **Length:** 14 in. (36 cm). **Wingspan:** 29 in. (74 cm).

AMERICAN AVOCET
Recurvirostra americana

A boldly black-and-white-patterned shorebird with deep rusty orange head and neck and long, thin, upturned bill. Underparts white. Legs powder blue. Immatures have same pattern but lack orange head and neck. A western species that has spread eastward and breeds on Gulf Coast. Birds also wander northward along coast to Virginia. **Length:** 18 in. (46 cm). **Wingspan:** 31 in. (79 cm).

AMERICAN OYSTERCATCHER
Haematopus palliatus

A large black-and-white shorebird with unmistakable bright orange bill. No other shorebird in the area has a similar bill. It is flattened like a knife blade and inserted into bivalves such as oysters to pry open shells. Juveniles similar to adults, but with more distinct feather edges on back. Ranges along Gulf Coast and East Coast as far as north New England. **Length:** 18 in. (46 cm). **Wingspan:** 32 in. (81 cm).

American Avocet landing. An adult avocet fading into winter (nonbreeding) plumage. Photographed at Fort Desoto, St. Petersburg, Florida.

N. PROCTOR

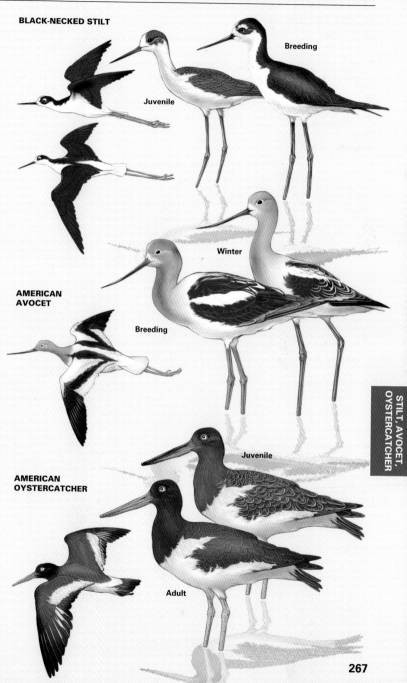

BLACK-NECKED STILT

Juvenile

Breeding

AMERICAN AVOCET

Winter

Breeding

AMERICAN OYSTERCATCHER

Juvenile

Adult

LESSER YELLOWLEGS
Tringa flavipes

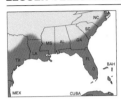

A more delicate-looking sandpiper than the Greater Yellowlegs. When seen alone, separation is difficult. Call a single "tsip" or double "tu-tu." Found across United States in migration. Wintering in Southeast from Virginia south to Florida and Gulf region. **Length:** 10.5 in. (25 cm). **Wingspan:** 25 in. (64 cm).

GREATER YELLOWLEGS
Tringa melanoleuca

Larger and not as delicate in proportions as the Lesser Yellowlegs. Upper part brown flecked with white. Underparts show heavier barring on flanks than the Lesser in breeding plumage. Often feeds running about with bill in water swinging from side to side. Call a loud "teu teu teu." Found across United States in migration. Winters in Southeast from Virginia south to Florida and Gulf region. **Length:** 14 in. (36 cm). **Wingspan:** 28 in. (71 cm).

WILLET
Catoptrophorus semipalmatus

Grayish brown with distinct back and underpart barring in breeding plumage. Very vocal, calling its name repeatedly: "pill-will-willet!" Favors shoreline and upper beach areas along Gulf and East Coasts north to Canada. **Length:** 15 in. (38 cm). **Wingspan:** 26 in. (66 cm).

Greater Yellowlegs, Sarasota, Florida. Note the heavier base of the bill than seen in the Lesser Yellowlegs, the relatively long bill, and the slight upturn of the bill near the tip.

N. PROCTOR

LESSER YELLOWLEGS

Delicate, straight bill with no upturn

Breeding

Winter

Bright yellow legs

GREATER YELLOWLEGS

Winter

Proportionately longer bill; usually shows a slight upturn

Bright yellow legs

Breeding

WILLET

Breeding

Winter

Heavy bill

The wing pattern is unmistakable

Gray legs

YELLOWLEGS, WILLET

269

SPOTTED SANDPIPER
Actitis macularia

A slim sandpiper that holds itself in a tilted forward position and bobs its body up and down as it walks. In breeding plumage, underparts heavily spotted and bill is bright yellow with dark tip. Ranges across United States, wintering on southern coasts from Carolinas through Florida and along Gulf Coast. **Length:** 7.5 in. (19 cm). **Wingspan:** 15 in. (38 cm).

PURPLE SANDPIPER
Calidris maritima

A stocky shorebird of windswept rocks and breakwaters of winter beaches. Fairly tame and approachable before it flutters off. Breeds in high Arctic. Winters on East Coast as far south as Carolinas occasionally to Florida and Gulf coasts. **Length:** 9 in. (23 cm). **Wingspan:** 17 in. (43 m).

RUDDY TURNSTONE
Arenaria interpres

A boldly marked shorebird with orange legs. Rich reddish chestnut back, black bib, and black facial markings in breeding plumage. Walks with a quick jerky stride and constantly moves the head looking for food. Another high Arctic breeder that winters along East and Gulf Coasts. Nonbreeding birds remain in these areas throughout the year. **Length:** 9.5 in. (24 cm). **Wingspan:** 21 in. (53 m).

A Purple Sandpiper in winter (nonbreeding) plumage on breakwater rocks, a typical view of a tough little bird. Note the very compact body form, short legs, and dark, decurved bill with yellow at the base.

N. PROCTOR

SPOTTED SANDPIPER

Flight pattern is stiff wingbeats followed by short glides

Breeding

Winter

Note the typical "tilted" posture and bobbing motion as it feeds

PURPLE SANDPIPER

Yellowish, decurved bill

The "purple" in the name comes from a reflective tint to the back feathers in breeding plumage

Breeding

Winter

Yellow-green legs

RUDDY TURNSTONE

The bold plumage pattern is unmistakable, winter or summer

Breeding

Winter

271

WHIMBREL   *Numenius phaeopus*

Large shorebird with striped crown and long, downcurved bill. Breeds near Hudson Bay, moving to East Coast in fall and returning north along coast in spring. Although many Whimbrels winter along southern US shores, others head to the Caribbean and into Central and South America. It may be these long-distance travelers that we see passing over wave tops well offshore. One of the few large shorebirds often seen far from land in migration. **Length:** 17 in. (43 cm). **Wingspan:** 32 in. (81 cm).

LONG-BILLED CURLEW   *Numenius americanus*

A large shorebird that is nearly unmistakable with its extremely long, downcurved bill. Back rich buffy brown, underparts light brown and unmarked. Crown brown and unmarked. In flight, underwing is rich cinnamon. The roughly similar Whimbrel is smaller and has a striped crown and a shorter bill. In flight, the Whimbrel has a distinctly pale rump, whereas the brown rump of the Long-Billed Curlew does not contrast with the back. Breeds on prairies and western sandhills and moves through coastal Carolina south to Florida and Gulf Coast. **Length:** 23 in. (58 cm). **Wingspan:** 35 in. (89 cm).

The Long-Billed Curlew is North America's largest member of the sandpiper family. The extremely long bill is used to probe mudflats and shallows for aquatic invertebrates.

N. PROCTOR

WHIMBREL

Whimbrel in flight

Grayer back than the Long-Billed Curlew

Gray legs

Curlew in flight

Extremely long bill

LONG-BILLED CURLEW

More buffy breast and more rufous back than the Whimbrel

Extremely long bill, even longer than that of the Whimbrel

273

KNOT, DUNLIN, DOWITCHERS

RED KNOT
Calidris canutus

A stocky, long-winged, short-billed shorebird that can be seen in both breeding and winter plumages as it passes through the area in migration. Breeds in the high Arctic and passes through coastal areas from Cape Hatteras to Florida and along Gulf coast. Small resident, nonbreeding populations can be found in isolated areas along west coast of Florida. Population is dropping due to the destruction of Horseshoe Crabs, on whose eggs it feeds in spring. **Length:** 11 in. (28 cm). **Wingspan:** 23 in. (58 cm).

DUNLIN
Calidris alpina

Rusty red back, black underbelly. Long bill curves down at tip. In nonbreeding plumage changes to nondescript brown-gray plumage on back, head, and chest with paler underparts. Bill is a key mark, black and curved downward at tip. Passes along East and Gulf Coasts in spring and fall migration. Nonbreeding young and older birds may remain on coasts throughout the year in small numbers. **Length:** 9 in. (23 cm). **Wingspan:** 17 in. (43 cm).

LONG-BILLED DOWITCHER
Limnodromus scolopaceus

SHORT-BILLED DOWITCHER
Limnodromus griseus

These two species have long perplexed many birders because of their extreme similarity and the complex subtleties of plumage that allow identification by experts. Best separated by flight call. Long-Billed: a single, high-pitched "kereep." Short-Billed: a rapid "tu-tu-tu." Bill length can overlap between the two species, so length alone is not much help even though the species names imply such. Both species winter along East and Gulf Coasts. **Length:** Both species, 11 in. (28 cm). **Wingspan:** Both species, 19 in. (48 cm).

Long-billed	Short-billed
Rufous breast in adult	Buff breast in adult
Rufous underbelly	White underbelly
Barring on neck	Spotting on neck
Grayer winter plumage	Browner winter plumage

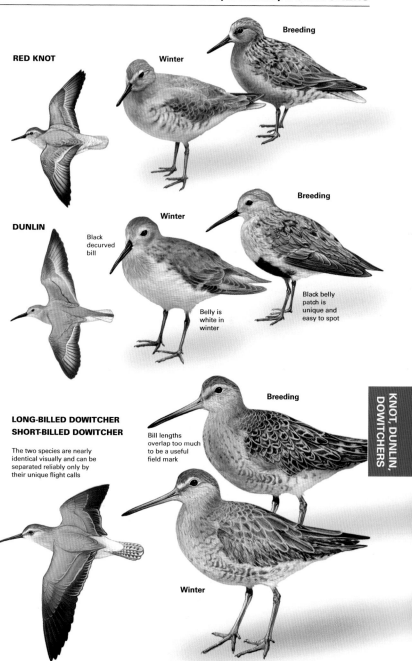

RED KNOT

Winter

Breeding

DUNLIN

Black decurved bill

Winter

Breeding

Belly is white in winter

Black belly patch is unique and easy to spot

LONG-BILLED DOWITCHER
SHORT-BILLED DOWITCHER

The two species are nearly identical visually and can be separated reliably only by their unique flight calls

Bill lengths overlap too much to be a useful field mark

Breeding

Winter

KNOT, DUNLIN, DOWITCHERS

275

WHITE-RUMPED SANDPIPER

Calidris fuscicollis

A small sandpiper with a reddish flush to cap and back in breeding plumage. In flight, rump pure white. In both winter and breeding plumage, note chevron marks on side below wings and onto flanks. Look for it in mixed flocks of other sandpipers of similar size, such as Semipalmated and Western Sandpipers. Breeds in the high Arctic and migrates through the eastern United States in spring and fall. Some populations winter as far south as Tierra del Fuego. **Length:** 7.5 in. (19 cm). **Wingspan:** 17 in. (43 cm).

SANDERLING

Calidris alba

Common shorebird of beaches along the East, West, and Gulf Coasts. Anyone who spends time at the beach is familiar with these small sandpipers chasing the waves in and out like tiny wind-up toys. In winter plumage, very white-looking with gray back and jet black legs. In breeding plumage, rich cinnamon red around head and neck. Often tame, allowing a close approach before taking flight and landing a short distance away. Nests in the high Arctic and seen along the coast in spring and fall migration. Many birds winter along southeastern and Gulf coasts. **Length:** 8 in. (20 cm). **Wingspan:** 17 in. (43 cm).

Sanderlings in August, in transition from breeding plumage to winter (nonbreeding) plumage. Note the patchy look of their faces, as the rich cinnamon of their breeding plumage transitions to light gray or white in winter. Corolla Beach, Outer Banks, North Carolina.

P. LYNCH

WHITE-RUMPED SANDPIPER

Bright white rump

Breeding

Chevron marks fade in winter, but are still present

Winter

SANDERLING

Breeding

Cinnamon red neck and breast

A very white-looking bird in winter plumage

Winter

SANDPIPERS

SANDPIPERS

LEAST SANDPIPER
Calidris minutilla

A small sandpiper that prefers muddy areas toward the rear of the beach and pool edges with grass rather than the sandy beach itself. In breeding plumage note yellow legs—Semipalmated and Western Sandpipers have black legs. In nonbreeding plumage, legs yellow-green. Even in nonbreeding plumage, crown and back are brownish rather than grayish. Adopts a crouched posture as it feeds. In spring migration ranges across the United States. Winters along coast from Carolinas south and all along Gulf Coast. **Length:** 6 in. (15 cm). **Wingspan:** 13 in. (33 cm).

SEMIPALMATED SANDPIPER
Calidris pusilla

The name implies partial webbing between the toes. Black legs. Overall much grayer than the Least Sandpiper. In breeding plumage, grayish brown back, cap, and sides. In winter plumage, gray with fairly dark cap and cheek. Bill black. Upper chest shows grayish band. Alert posture. A migrant across the eastern two-thirds of the United States from breeding grounds. Common on the East and Gulf Coasts, often in large flocks. **Length:** 6.25 in. (16 cm). **Wingspan:** 14 in. (36 cm).

WESTERN SANDPIPER
Calidris mauri

Easily confused with the Semipalmated Sandpiper. Note longer black bill, which curves downward throughout its length. In breeding plumage, rufous cap and shoulder area. Even in grayish winter plumage there is often rufous on shoulder. Winter plumage also shows lighter gray to cap and face area, giving it a "white-faced" appearance. When mixed in with Semipalmated Sandpiper flocks, it can be a challenge to pick out which is which. Mainly a winter visitor to the East and Gulf Coasts from its normal western migration route. **Length:** 6.5 in. (17 cm). **Wingspan:** 14 in. (36 cm).

LEAST SANDPIPER

Breeding

Brownish breast

Winter

Yellow legs

Yellow-green legs

SEMIPALMATED SANDPIPER

Breeding

Winter

WESTERN SANDPIPER

Drooping bill

Breeding

Very light face in winter plumage

Winter

SANDPIPERS

SKUAS

These massive birds are the top aerial predators of the Atlantic Ocean. Superficially similar to large gulls, they are very different in flight pattern and behavior. Bulky bodied, skuas are powerful fliers with deep, stiff wingbeats. Like jaegers, they nest on land near seabird colonies that afford them plenty of food for raising their young, then they spend the rest of the year at sea. Skuas do not flock—usually a single bird will arrive near a boat and harass all the other birds (including jaegers) for their food.

GREAT SKUA   *Catharacta skua*

Massive and barrel-shaped, with large head and short tail tapering to triangular end. Chocolate brown overall with large white patches at base of outer wing feathers. Flight is slow and steady, its power deceptive as this skua rapidly closes in on an escaping bird. A powerful hunter that defers to no other seabird. Very opportunistic; if it's edible, this bird will eat it. Skuas often chase Northern Gannets and flip them over in the air by hooking their wingtips under the gannets' and pushing upward, causing the gannets to crash into the water and disgorge their food. **Length:** 23 in. (58 cm). **Wingspan:** 55 in. (1.4 m).

SOUTH POLAR SKUA   *Catharacta maccormicki*

Large and bulky, similar in shape to Great Skua. Two color phases occur, dark and blond (light). Dark-phase body feather color more grayish than brown overall. Blond phase easily identified by very light head and shoulders. Both phases show large white outer wing patches. This skua is most usefully separated from Great by bronze-blond nape, visible in most dark-phase adults that enter our area and absent in all Great Skuas. South Polar Skuas in our area are nonbreeding birds spending the summer in our area, always well offshore in deeper waters. **Length:** 21 in. (53 cm). **Wingspan:** 52 in. (1.3 m).

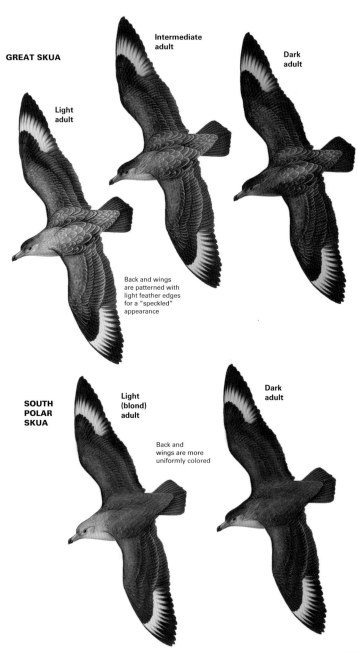

GREAT SKUA

Intermediate adult

Dark adult

Light adult

Back and wings are patterned with light feather edges for a "speckled" appearance

SOUTH POLAR SKUA

Light (blond) adult

Dark adult

Back and wings are more uniformly colored

SKUAS

JAEGERS

The "raptors" of the open ocean. Powerful and swift, these aggressive birds harass all species of oceanic birds. Three species of jaegers (German for "hunter") occur in our waters. More falcon- or hawklike than gull-like in appearance, jaegers are often recognized first by their deep, powerful wingbeats and aggressive behavior. Separation of the three species can be difficult, especially in young or dark-phase birds.

POMARINE JAEGER   *Stercorarius pomarinus*

A heavy-chested, round-bodied jaeger. Two color phases, light and dark. Spatulate, twisted tail feathers make identification easy in breeding plumage. Young Pomarines, however, have small tail feathers that may look like those of the smaller Parasitic Jaeger. Less white on outer wing feathers than in other jaegers; often only feather shafts show white. Distinct chest band in light phase. **Length:** 22 in. (56 cm). **Wingspan:** 48 in. (1.2 m).

PARASITIC JAEGER   *Stercorarius parasiticus*

Two color phases, light and dark. Central two tail feathers stick out, forming a sharp ventral point, in full adult plumage. Both color phases have considerable white in outer wing feathers. Light phase has light gray sides that blend into white belly and lacks distinctive barring of dark phase. **Length:** 18 in. (46 cm). **Wingspan:** 42 in. (1.1 m).

LONG-TAILED JAEGER   *Stercorarius longicaudus*

The trimmest of the three jaegers. In breeding plumage, the long (7 in., 18 cm) central tail feathers taper the body's appearance, almost suggesting a large, dark tern. No chest band, and very little white in outer wing feathers. The rarest of the jaegers off the Atlantic Coast, appearing only sporadically in migration, usually far offshore. **Length:** 22 in. (56 cm). **Wingspan:** 40 in. (1 m).

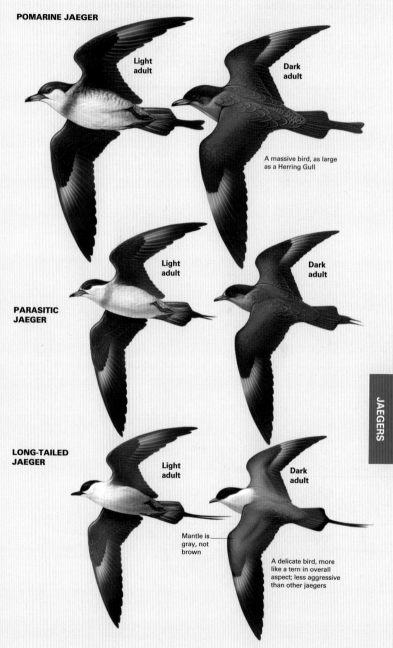

POMARINE JAEGER

Light adult

Dark adult

A massive bird, as large as a Herring Gull

PARASITIC JAEGER

Light adult

Dark adult

LONG-TAILED JAEGER

Light adult

Dark adult

Mantle is gray, not brown

A delicate bird, more like a tern in overall aspect; less aggressive than other jaegers

JAEGERS

GREAT BLACK-BACKED GULL

Larus marinus

A common bird of harbor areas and shorelines. Of all the large inshore gulls, this species will follow boats out to sea the farthest. The Great Black-Backed Gull is gradually extending its range down the eastern seaboard toward Florida.

Description: The largest gull of the shoreline. Jet-black back contrasts sharply with pure white underparts. White border shows from wingtips along rear edge of wings when in flight. Heavy yellow bill with red spot. Flesh-colored legs. First-year immatures can usually be told by massive size and by pale head and rump contrasting with brown back and underparts. **Length:** 30 in. (76 cm). **Wingspan:** 65 in. (1.7 m).

Habits: Aggressive and territorial with other birds, but will often nest peacefully in mixed-species colonies with other gulls. Found throughout northeast Atlantic Coast all year north of North Carolina and in winter south to northern Florida. Casual to Gulf Coast.

Similar species: Superficially similar to the Herring Gull in plumage and overall size, but note the much darker back, massive bill, and larger, bulkier profile. The uncommon Lesser Black-Backed Gull is similar but not covered in this book except as seen in flight (see pages 299 and 301).

Great Black-Backed Gull
First winter

Massive dark bill

"Checkered" effect of high-contrast plumage pattern

Herring Gull
First winter

Dull brown mantle lacks contrast

Smaller bill, light at the base

Length: 30 in.

Length: 25 in.

Comparison of young Herring and Great Black-Backed Gulls, which are often seen in mixed flocks on Atlantic Coast beaches.

Third winter

The stark black-and-white contrast of the adult is often the best field mark at a distance

Adult winter

The largest gull in North America

First winter

Second winter

Second winter

Heavy bill at all ages

First winter

Adult breeding

Pale pink legs at all ages

285

HERRING GULL

Larus argentatus

Description: An abundant bird—the gull most people think of when they think of a "seagull." Large, with gray back and black wingtips with white spots. White head and underparts. Flesh-colored legs. Yellow bill with blood-red spot on lower mandible. Yellow eye. In winter plumage, head is streaked brown with a dark eye line, giving the face a "mean" appearance. First-year immatures are chocolate brown with lighter speckles. Herring Gulls reach full adult plumage after four years. Plumage may be distinguished by year of age until adult plumage is reached. **Length:** 25 in. (64 cm). **Wingspan:** 58 in. (1.5 m).

Habits: An aggressive, opportunistic bird, readily adapting to both natural and man-made environments from well inland to miles from shore at sea. Will follow fishing boats well away from land. On the coast will often pick up shellfish and crabs and drop them from a height to crack their shells. Ranges along the entire East Coast in winter and from Maritime Canada to the Carolinas year-round.

Similar species: The similar-looking Ring-Billed Gull is smaller, with a more delicate bill. See also the comparison of first-year birds on page 288.

Great Black-Backed Gull
Length: 30 in.

Herring Gull
Length: 25 in.

Ring-Billed Gull
Length: 18 in.

The larger white-headed gulls are superficially similar but separate distinctly by size. The Great Black-Backed Gull is a much more massive bird than the Ring-Billed Gull.

First winter

Dull brown mantle and lighter brown breast lacks contrast

Adult

Second winter

Gray central mantle and partially gray wings

Tail very dark compared to the similar second-winter Ring-Billed Gull

Third winter

Similar to adult winter, but with a darker head and tail

Second winter

Adult breeding

First winter

Pink legs at all ages

287

RING-BILLED GULL

Larus delawarensis

Description: A sleek, medium-sized gull of harbors, shorelines, and shopping center parking lots. Easily identified by distinct ring around bill. Very common in our area during the colder months. Gray back. Greenish yellow legs. Black wingtips spotted with white; black color extends along fore edge almost to wing bend. Reaches full adult plumage after three years. **Length:** 18 in. (46 cm). **Wingspan:** 48 in. (1.2 m).

Habits: A very flexible, opportunistic species that has done very well in adapting to human development of the coastline. Mixes with other gulls in harbors and in large flocks resting on breakwaters and sandy shores. Will follow boats, hanging in the wind just astern and looking for handouts. Does not follow boats far offshore. Ranges along the entire Atlantic and Gulf Coasts in the colder months. Breeds mostly in north-central Canada in the summer months.

Similar species: The Herring Gull is the most similar (see below and pages 286–287). If you get a chance to see the two species side by side, note the much smaller, lighter body and more delicate features of the Ring-Billed Gull.

Herring Gull
First winter
Length: 25 in.

Mottled brown back

Ring-Billed Gull
First winter
Length: 18 in.

Diffuse dark bill tip

Much gray in the back

Well-defined black tip

Comparison of first-winter Herring and Ring-Billed Gulls.

First winter

A more contrasting pattern on the mantle than in first-winter Herring Gulls

Adult breeding

Second winter

Gray mantle, showing minimal or no brown remnants

Tail is much lighter than in the similar second-winter Herring Gull

First winter

Relatively small bill at all ages

Adult winter

Bill ring

Adult breeding

Yellow legs at all ages

LAUGHING GULL

Larus atricilla

Description: A trim gull with a black head and deep gray back. Blood-red bill and legs. No white in wingtips. Note broken white ring around eye. In winter plumage, hood fades to dark patch at back of head. Immature shows a black band at end of tail feathers. 17 in. (43 cm). **Wingspan:** 40 in. (1 m).

Habits: This abundant gull's laughing call is a familiar sound from the mid-Atlantic Coast southward, and the Laughing Gull is becoming more common in the Northeast as the climate warms. Will follow inshore boats, "hanging" above the stern in search of handouts. Ranges along entire East Coast in the warmer months and from Cape Hatteras south year-round.

Similar species: See the comparisons to the Franklin's Gull on this page and on pages 292–293 for details on these very similar species, which overlap only on the Gulf Coast west of and around New Orleans. Bonaparte's Gull also has a black head in breeding plumage but is much smaller and more ternlike, and Bonaparte's Gull rarely mixes with the Laughing Gull in flocks.

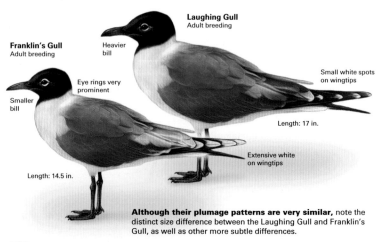

Laughing Gull
Adult breeding

Franklin's Gull
Adult breeding

Heavier bill

Eye rings very prominent

Smaller bill

Small white spots on wingtips

Length: 17 in.

Extensive white on wingtips

Length: 14.5 in.

Although their plumage patterns are very similar, note the distinct size difference between the Laughing Gull and Franklin's Gull, as well as other more subtle differences.

First
winter

Adult
breeding

Shows
brown
in wing
mantle

Heavy
terminal
band

Mantle
entirely
gray

Second
winter

Very light
terminal
band

Wingtips
black

First
winter

Adult
winter

Dark
bill

Adult breeding

Legs dark gray

Dark
red bill

Legs dark gray

Legs dark
red in
breeding
plumage

FRANKLIN'S GULL

Larus pipixcan

Description: A small gull closely related to the larger Laughing Gull. Very similar in all respects to the Laughing Gull, but smaller, with more prominent rings around eyes and more white on wingtips. Bill is more slender and delicate than the Laughing Gull's.
Length: 14.5 in. (37 cm). **Wingspan:** 36 in. (91 m).

Habits: Like most gulls, Franklin's Gull is omnivorous and will eat virtually any small animal it can catch or scavenge. On its breeding range the Franklin's eats mostly small insects, fish, and amphibians, as well as earthworms and grubs exposed on farmland. Franklin's Gull breeds in the north-central United States and Canada, where it frequents farmlands and prairies. Seen in our area only in migration, and then only on Louisiana and Texas coasts during spring and fall. Winters on Pacific Coast from Central America to Chile. Rare on East Coast or elsewhere on Gulf Coast.

Similar species: Very similar to the Laughing Gull, but the two species are seen together only during migration, and only on western Gulf Coast. See the comparison figures immediately below and on pages 290–291 for visual keys to separating Laughing Gull and Franklin's Gull.

Laughing Gull
First winter

Franklin's Gull
First winter

Heavier bill

Eye rings more prominent

Smaller bill

Length: 17 in.

Length: 14.5 in.

In young Franklin's Gulls, look for the darker face plumage, and note the smaller, more delicate bill than in young Laughing Gulls. Franklin's Gulls are also smaller than Laughing Gulls.

First winter

Shows brown in wing mantle

Light terminal band

Mantle entirely gray

Second winter

White tail

At all ages the bill is proportionately smaller than in the Laughing Gull

Adult breeding

Wingtips show extensive white

First winter

Legs dark gray

Adult winter

Legs dark gray

Prominent eye ring

Red bill

Adult breeding

Extensive white areas on wingtips

Legs red in breeding plumage

FRANKLIN'S GULL

BONAPARTE'S GULL

Larus philadelphia

Description: A small, almost ternlike gull. Very buoyant when sitting on the water. Black head in breeding plumage. Wings in all plumages show a distinct white wedge on outer edge. Blood-red bill, deep pink legs. In winter plumage, black hood is reduced to a black smudge behind eye. **Length:** 13 in. (33 cm). **Wingspan:** 33 in. (84 cm).

Habits: A fall-through-spring visitor to Atlantic and Gulf Coasts. At times found well offshore, where it mixes with true oceanic birds. Bonaparte's is a small gull that doesn't often mix with the larger gull species, preferring to stand apart from mixed flocks on beaches and sandbars—a useful tip for spotting small flocks of Bonaparte's Gulls. Ranges in winter throughout Atlantic and Gulf Coasts and lower Mississippi River Valley.

Similar species: The small size and delicate bill separate it from the more common and much larger Laughing Gull. On the mid-Atlantic Coast the Bonaparte's range overlaps with two other small gulls that are rare in our southeastern and Gulf area and thus are not covered here: the Little Gull and the Black-Headed Gull.

Laughing Gull
Winter (nonbreeding)
Length: 17 in.

Bonaparte's Gull
Winter
Length: 13 in.

Comparison of first-winter Bonaparte's and Laughing Gulls.

White wedge near the wingtips

Adult breeding

Dark tips with a light band

First winter

Dark wing bars

Tail band

First winter

Cheek spot

Adult winter

Delicate bill

Prominent eye ring

Adult breeding

Red legs

BONAPARTE'S GULL

Two gull species that are true ocean wanderers off the eastern seaboard. After breeding in the Arctic and along the far northern coasts, they wander out to sea to spend the winter. Severe storms on the Atlantic occasionally drive these birds inshore with coastline gull flocks, but normally you'll never see these truly pelagic species within sight of land.

BLACK-LEGGED KITTIWAKE   *Rissa tridactyla*

A small gull with a dovelike head. Wingtips have a dipped-in-ink pattern that shows a clear edge contrast with gray back. Yellow, unmarked bill. Immature shows a striking dark M pattern across back and upper wings in flight. Black tail edge; distinct black collar on neck.

A gull that loves storm-tossed seas, often appearing in incredible numbers far offshore. Flies in circular patterns, rising in the air then dipping down to the surface level and riding up again. These large, looping patterns are diagnostic from a great distance. Follows boats until land is in sight. **Length:** 17 in. (43 cm). **Wingspan:** 35 in. (89 cm).

SABINE'S GULL   *Xema sabini*

A beautifully patterned gull, the only gull with a forked tail. Note deep gray hood, black bill with yellow tip, and striking back pattern formed by black outer wing feathers contrasting with wedge of white on inner wing and deep gray back. Immature has a unique scalelike pattern on its gray-brown back.

A gull of the deep ocean. Rarely seen inshore except after severe storms. Buoyant when seen sitting on the water, its tail held upward. A strong flyer. Will come to chumming or activity of birds around a boat. Rarely seen; a standout sighting on any offshore trip. **Length:** 14 in. (36 cm). **Wingspan:** 33 in. (84 cm).

BLACK-LEGGED KITTIWAKE

First winter

First winter

Adult breeding

Adult breeding

SABINE'S GULL

First winter

Bill dark

The only gull in our area with a forked tail

Short bill with yellow tip

First summer

Adult winter

Adult breeding

Bold wing pattern in all plumages

ADULT GULLS IN FLIGHT

All species are shown to scale; some less common gull species are included on this plate but not in the main text

ICELAND GULL

Larus glaucoides

Wingspan: 54 in.

GLAUCOUS GULL

Larus hyperboreus

60 in.

These two very similar large "white-winged" gull species are rare but regular visitors south of Cape Hatteras. Birding experts in the Carolinas occasionally spot them in large flocks of mixed gull species.

BONAPARTE'S GULL

33 in.

FRANKLIN'S GULL

Rarely seen outside of Texas and Louisiana, and then only in spring and fall migration. Winters off the coast of South America.

36 in.

LITTLE GULL

Larus minutus

24 in.

A tiny, ternlike gull that is uncommon in its normal northeastern coastal range and rare and local south of Virginia

LAUGHING GULL

40 in.

**LESSER
BLACK-BACKED
GULL**

Larus fuscus

A European species
that now appears
uncommonly but
regularly along the East
and Gulf Coasts. Unusual,
and difficult to separate
from the very similar
Great Black-Backed Gull.
Note the more gray
mantle, and somewhat
smaller size.

54 in.

**GREAT
BLACK-BACKED
GULL**

65 in.

HERRING GULL

58 in.

**RING-BILLED
GULL**

48 in.

**BLACK-LEGGED
KITTIWAKE**

35 in.

**SABINE'S
GULL**

33 in.

These are uncommon pelagic
gulls rarely seen near land

ADULT GULLS
IN FLIGHT

299

IMMATURE GULLS IN FLIGHT

All species are shown to scale; some less common gull species are included on this plate but not in the main text

ICELAND GULL

Wingspan: 54 in.

GLAUCOUS GULL

60 in.

These two very similar large "white-winged" gull species are rare but regular visitors south of Cape Hatteras. Birding experts in the Carolinas occasionally spot them in large flocks of mixed gull species.

BONAPARTE'S GULL

33 in.

FRANKLIN'S GULL

Rarely seen outside of Texas and Louisiana, and then only in spring and fall migration. Winters off the coast of South America.

36 in.

LITTLE GULL
Larus minutus

24 in.

A tiny, ternlike gull that is uncommon in its normal northeastern coastal range and rare and local south of Virginia

LAUGHING GULL

40 in.

LESSER BLACK-BACKED GULL

Larus fuscus

A European species that now appears uncommonly but regularly along the East and Gulf Coasts. Unusual, and difficult to separate from the very similar Great Black-backed Gull. Note the grayer mantle and somewhat smaller size.

54 in.

GREAT BLACK-BACKED GULL

65 in.

HERRING GULL

58 in.

RING-BILLED GULL

48 in.

BLACK-LEGGED KITTIWAKE

35 in.

SABINE'S GULL

33 in.

These are uncommon pelagic gulls rarely seen near land

IMMATURE GULLS IN FLIGHT

COMMON TERN, FORSTER'S TERN

Most tern species are inshore birds that rarely wander more than a mile offshore. Terns are most often seen feeding in coastal waters, particularly near their breeding colonies.

COMMON TERN   *Sterna hirundo*

Gray back and upper wings, white belly and underwings. Black cap extends down nape. Orange-red bill with black tip. In flight, gray wedge at center of back and darker wingtips. In winter plumage, white forehead and front half of crown. Immatures roughly follow adult winter pattern, with dark leading edge on upper wing. Very young birds are flecked with fine brown spotting. **Length:** 14 in. (36 cm). **Wingspan:** 30 in. (76 cm).

FORSTER'S TERN   *Sterna forsteri*

Similar to the Common Tern, but with pale wingtips and more orange base of bill. Wintering terns are easier to identify: Forster's Tern has a heavy black eyeline, an all-white crown, and a gray nape. Other winter terns have a half-black crown and a black nape. **Length:** 14 in. (36 cm). **Wingspan:** 31 in. (79 cm).

Common Tern and chick, Stewart B. McKinney National Wildlife Refuge, Falkner Island, off Guilford, Connecticut. Note how well the chick is camouflaged. *Please stay out of tern colonies during breeding season.* The disturbed adults are unable to shelter their young against the heat and sun, and the nests and chicks are so hard to spot that they are very vulnerable to a misstep.

COMMON TERN

Breeding

Winter

In all plumages the Common Tern has a white breast and a clean white face and cheeks

At rest, the tail extends only to the tip of the folded wings

FORSTER'S TERN

Breeding

Winter

Wingtips are very pale

TERNS

ROSEATE TERN, LEAST TERN

Stay out of coastal and island tern colonies! Most tern populations are falling as these birds lose their breeding areas on beaches and coastal islands to human bathers and boaters. Terns are very defensive of their nesting sites. Disturbing their breeding areas causes high mortality, as chicks get lost and eggs get stepped on. Habitat loss and human disturbance of nesting sites have devastated these two species of terns in particular.

ROSEATE TERN
Sterna dougallii

Uncommon. Pale gray to almost white. In breeding plumage, breast has a faint pink cast. Black bill with a deep red base. Long tail feathers form a deeply forked tail. In flight, wings are a clear, very pale gray. Outermost two primary feathers dark gray. Feeds in inshore waters. An endangered species, common only around its few remaining breeding colonies. **Length:** 15 in. (38 cm). **Wingspan:** 29 in. (73 cm).

LEAST TERN
Sterna antillarum

The smallest tern. White forehead mark is present even in summer. The only tern with a yellow bill, tipped with black. Immatures have especially noticeable dark leading edges on upper wings. An inshore bird; nests on beaches and, in some areas, on flat rooftops near the shoreline. **Length:** 9 in. (23 cm). **Wingspan:** 20 in. (51 cm).

Roseate Terns, Common Terns, and Forster's Terns have very similar plumage patterns, but the Roseate always has a lighter-bodied, more delicate look than its close relatives, and it has a more sharply pointed dark bill. This Roseate Tern was photographed in a breeding colony on Falkner Island, Connecticut. Note the bright red base of the bill in breeding plumage.

P. LYNCH

ROSEATE TERN

Tail is longer and lighter than in other terns

The Roseate Tern is distinctly lighter than other small terns and has a very buoyant flight pattern

Breeding

Winter

Wingtips are gray, not black

The slightly pink cast to the belly is rarely visible except close-up and even then is not a reliable field mark

LEAST TERN

Much smaller than other terns and very pale on the back and wings

Breeding

Winter

Yellow bill is unique

TERNS

305

GULL-BILLED TERN
Sterna nilotica

Pale gray with black cap and very heavy black bill. Stout, stocky body contrasts with the sleeker Common and Forster's Terns. Black legs and feet. In winter plumage, head almost completely white. An inshore bird of salt marshes and bays. The Gull-Billed Tern was hard-hit by the early twentieth-century fashion for feathers in women's hats, and their populations are just recovering after nearly a century of very slow growth. **Length:** 14 in. (36 cm). **Wingspan:** 34 in. (86 cm).

SANDWICH TERN
Sterna sandvicensis

A sleek, pale tern. Long black bill with yellow tip (yellow tip may not show in young birds). Black legs. Black cap with slight crest. White forehead in winter plumage. Very similar in size and body pattern to the Common Tern, but with much lighter wings and wingtips. Often you'll see mixed flocks of Sandwich and Royal Terns on beaches and sandbars. Rarely wanders far from shore but is a strictly marine bird, not often seen near freshwater shores. **Length:** 15 in. (38 cm). **Wingspan:** 34 in. (86 cm).

A threat display between Sandwich Terns near their breeding colony. In minor disputes over turf and territory, one or both birds will rear up in an aggressive-upright posture, stiffly holding the wings down and forward, as a warning to other Sandwich Terns to "back off" and stay away.

N. PROCTOR

Tail is short and not deeply notched

GULL-BILLED TERN

Breeding

Winter

Outer wings are long and graceful and always pale

Wingtips are gray, not black

SANDWICH TERN

Breeding

Winter

Long, thin bill with yellow tip

TERNS

LARGE TERNS

Large terns are studied to best advantage along the coastline, where the birds stand on pilings or on the beach. All are southern species that wander north in summer. The Royal Tern is the species most likely to be seen well offshore.

ROYAL TERN
Sterna maxima

A large, sleek tern. Black cap with slight crest. Long orange-yellow bill. Pale underwings. Deeply forked tail. Unstreaked white forehead in winter plumage. Aggregates into large roosting flocks. A plunge-diver that feeds along inshore waters. Will sometimes follow ships at sea. **Length:** 20 in. (51 cm). **Wingspan:** 41 in. (1 m).

CASPIAN TERN
Sterna caspia

A large, black-capped, heavy-billed tern nearly the size of a Herring Gull. Bright orange-red bill, with a tiny yellow area at tip. Wingtips dark below and lighter above, a reverse of the usual tern pattern. Shallowly forked tail. White-streaked forehead in winter plumage. Strictly a coastal species; does not wander far offshore. **Length:** 21 in. (53 cm). **Wingspan:** 50 in. (1.3 m).

Royal Terns in winter (nonbreeding) plumage, Fort Jefferson, Dry Tortugas National Park.

P. LYNCH

ROYAL TERN

Longer tail than in the Caspian Tern

Breeding

Winter

Light wingtips above are dark when seen from below

Shaggy crest

Bill is *yellow-orange*, not red-orange as in the Caspian Tern

CASPIAN TERN

Breeding

Winter

Slight crest

Heavy bill is red-orange with a dark area near the tip

TROPICAL TERNS

SOOTY TERN
Sterna fuscata

A large tern usually seen well offshore. Black upperparts contrast sharply with bright white underparts. Dusky gray underwings. The long wings are held high when gliding, with a characteristic sharp bend at the wrist. White-edged tail is deeply forked. White forehead does not extend past eye. Cannot land on water due to its poor waterproofing. Feeds by plucking prey from water surface with its long bill. A tern most likely to be seen in our area after hurricanes and tropical storms in the north. **Length:** 16 in. (41 cm). **Wingspan:** 32 in. (81 cm).

BRIDLED TERN
Sterna anaethetus

A rare tern from the Bahamas and West Indies that is sometimes blown into our area by hurricanes. Similar to the Sooty Tern but lighter colored on the back, with a white collar at the nape. **Length:** 15 in. (38 cm). **Wingspan:** 30 in. (76 cm).

SOOTY TERN

Dark "sooty" head and breast in the first few months of life

Juvenile

Winter

White usually confined to the forehead

Black nape

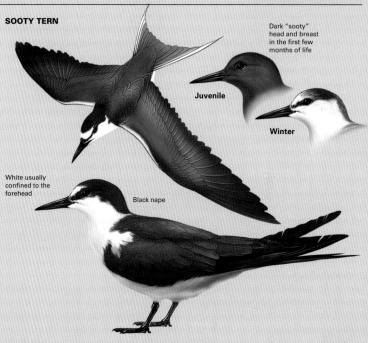

BRIDLED TERN

Lighter tail than in the Sooty Tern

White collar

Juvenile

Winter

White extends over eye

Back and wings are lighter than the Sooty Tern's

311

DARK TERNS

BROWN NODDY

Anous stolidus

A brown tern with grayish white crown. Long, thin black bill. Long, wedge-shaped tail unique among terns. Unlike the Sooty Tern, doesn't wander north; rarely seen in our area except after hurricanes. Most Brown Noddies seen are immatures blown from nesting colonies in the Dry Tortugas, off the Florida coast. **Length:** 15 in. (38 cm). **Wingspan:** 32 in. (81 cm).

BLACK NODDY

Anous minutus

A very rare visitor from the Caribbean, sometimes blown north along the Atlantic Coast by hurricanes and other storms. The Black Noddy is a smaller, more delicate tern than the Brown Noddy, with a longer, thinner bill. **Length:** 13 in. (33 cm). **Wingspan:** 30 in. (76 cm).

Florida's isolated Dry Tortugas National Park, 70 miles west of Key West, is one of the few places in the United States that you are likely to encounter noddies on land, particularly the rare Black Noddy. Bush Key (top photo, background of bottom photo) is a nesting site for large numbers of Brown Noddies, Sooty Terns, and Magnificent Frigatebirds.

P. LYNCH

BROWN NODDY

Juvenile

First summer

Cap is more diffuse than the Black Noddy's

BLACK NODDY

A sharply defined white or nearly white cap in all plumages

Adult (all year)

Juvenile

Darker than the Brown Noddy

Longer, more delicate bill than in the Brown Noddy

BLACK TERN   *Chlidonias niger*

A fall migrant along Atlantic Coast, usually seen in winter plumage. Dark gray back and upper wings; much darker than other terns. All-black bill. Very dark red to black legs and feet. **Length:** 10 in. (25 cm). **Wingspan:** 24 in. (61 cm).

BLACK SKIMMER   *Rynchops niger*

Unmistakable. A medium-sized relative of gulls and terns, with a unique long lower mandible used to "skim" surface waters for small fish. Dark, almost black back; white underparts. Huge bright-red bill with black tip. Most often noticed skimming across the surface of shallow coastal waters with its long lower bill slicing the water. Rarely wanders far from shore but often nests on offshore islands among or near tern colonies. **Length:** 18 in. (46 cm). **Wingspan:** 44 in. (1.1 m).

Black Skimmers rest in a unique way. The bird on the left isn't sick or injured. Perhaps because of the weight of their bills, skimmers will often stretch out across the sand with their knife-thin bills resting on the ground.

N. PROCTOR

BLACK TERN

Winter

Breeding

Winter

Adult breeding

BLACK SKIMMER

Winter

The nape of the
neck becomes white
or gray in winter

Immature

A brown, mottled
echo of the adult
plumage

Adult breeding

ADULT TERNS IN FLIGHT

Wingspan:
31 in.

COMMON TERN

FORSTER'S TERN

30 in.

GULL-BILLED TERN

34 in.

SANDWICH TERN

34 in.

ROYAL TERN

41 in.

LEAST TERN

20 in.

All species are
shown to scale

CASPIAN TERN

50 in.

ROSEATE
TERN

29 in.

BLACK
NODDY

30 in.

BROWN
NODDY

32 in.

BRIDLED
TERN

30 in.

SOOTY
TERN

32 in.

BLACK
SKIMMER

BLACK
TERN

24 in.

44 in.

ADULT TERNS
IN FLIGHT

317

TERNS—SIZE COMPARISON

BLACK TERN 10 in.

LEAST TERN Length: 9 in.

ROSEATE TERN 15 in.

BLACK NODDY 13 in.

COMMON TERN 14 in.

FORSTER'S TERN 14 in.

GULL-BILLED TERN 14 in.

BRIDLED TERN 15 in.

SANDWICH
TERN

15 in.

BROWN
NODDY

15 in.

SOOTY
TERN

16 in.

BLACK
SKIMMER

18 in.

CASPIAN
TERN

21 in.

ROYAL
TERN

20 in.

ADULT TERNS
STANDING

All species are shown to scale

319

FIN WHALE

Balaenoptera physalus

Description: A huge whale, dark gray to black on back and sides, with light gray or white underparts. Lower jaw asymmetrically colored: left side dark, right side light. Flippers and flukes small in comparison to body. At close range look for subtle chevron pattern of streaks from behind eyes to midline of back. Small, falcate dorsal fin sits far to rear, about three-fourths of the way back from jaw tip. **Red List – Endangered**

Habits: A sleek, fast swimmer despite its size. May approach drifting or very quiet boats, but is indifferent to most vessels and apparently shy of engine noises. Sometimes performs low lunges across the water's surface while pursuing schools of fish; rarely breaches. Typical surface behavior is two to five blows, followed by a dive of five to ten minutes or more. Rarely rolls out its tail when diving. As their tails sweep up through the water, submerged Fin Whales often leave huge circular "footprints," formed by upwelling water, on the surface.

Range: Fin Whales are most common along the coast from Cape Hatteras north to the Canadian Arctic in the warmer months, but they can occur anywhere along the East Coast, albeit in small numbers. Fin whales are very rarely seen in the Gulf of Mexico. Populations drift south in the winter months, and Fins are believed to calve off the mid-Atlantic Coast, their breeding and calving behavior is not well understood.

Similar species: The Sei Whale is smaller, has small oblong spots across the back, and has jaws that are dark on both sides. South of Virginia, Bryde's Whale has three ridges on the head and is much smaller than most Fin Whales.

Size: 30–70 ft. (9.1–21.3 m); average adult length 50–60 ft. (15.2–18.3 m).

BLOW PROFILE
(as seen from behind)

Fin blows are tall and narrow, typically up to 20 ft. (6 m) high. Noise of blows can be heard long distances across open water.

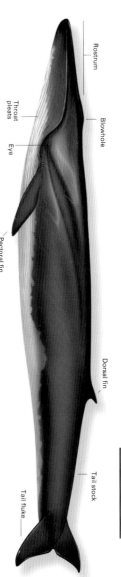

Rostrum

Throat pleats

Blowhole

Eye

Pectoral fin

Dorsal fin

Tail stock

Tail fluke

A fast, powerful swimmer that doesn't typically linger at the surface

SURFACE PROFILE

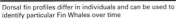

Dorsal fin profiles differ in individuals and can be used to identify particular Fin Whales over time

RIGHT = WHITE

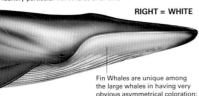

Fin Whales are unique among the large whales in having very obvious asymmetrical coloration: the right jaw is always bright white or light gray

Splashguard

TOP VIEW OF HEAD

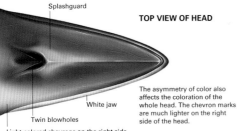

The asymmetry of color also affects the coloration of the whole head. The chevron marks are much lighter on the right side of the head.

White jaw

Twin blowholes

Light-colored chevrons on the right side

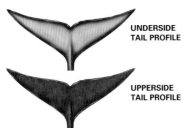

UNDERSIDE TAIL PROFILE

UPPERSIDE TAIL PROFILE

FIN WHALE

MINKE WHALE

Balaenoptera acutorostrata

Description: A small, fast-moving whale with a sharp snout. Its quickness and small size are often more suggestive of dolphins than any of the larger rorquals. Blue-gray, with distinctive light chevron marks on back. Belly light gray to white. Pectoral fins have white bands that are often visible under the water at close range. Falcate dorsal fin very similar to a dolphin, but positioned well to the rear of the back. **Red List – Near threatened**

Habits: An inquisitive whale that approaches boats almost in the manner of dolphins. Often swims under or near moving boats but is generally shy and difficult to approach. The Minke's small size makes it easy to miss as it approaches boats, and Minkes often seem to appear suddenly from nowhere even when many whale watchers are aboard. Minkes breach more often than other rorquals; even so, breaching is infrequent. Typical surface behavior is five to seven nearly invisible blows, followed by a dive of three to eight minutes. The sharp snout (rostrum) often pokes a few feet out of the water as the whale surfaces.

Range: One of the most widely distributed whale species, but not common in our area from Cape Hatteras south and extremely rare in the Gulf of Mexico. Minkes are mostly likely to be encountered in the colder months, moving well offshore of Cape Hatteras and the Carolinas.

Similar species: From a distance, may look like a lone dolphin or small toothed whale. Look for the sharp, dark snout as the Minke surfaces; it is unlike that of any dolphin or small toothed whale.

Size: 12–30 ft. (3.7–9.1 m); average adult length 18–22 ft. (5.5–6.7 m).

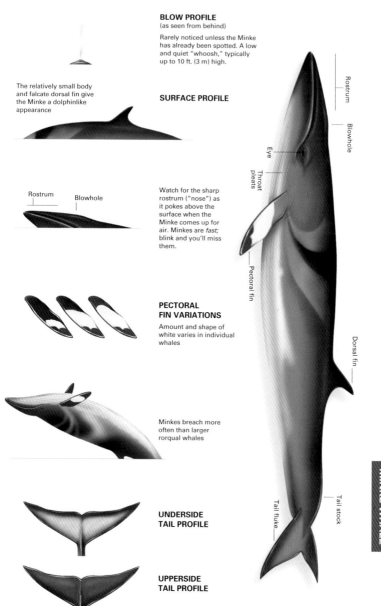

BLOW PROFILE
(as seen from behind)

Rarely noticed unless the Minke has already been spotted. A low and quiet "whoosh," typically up to 10 ft. (3 m) high.

The relatively small body and falcate dorsal fin give the Minke a dolphinlike appearance

SURFACE PROFILE

Rostrum   Blowhole

Watch for the sharp rostrum ("nose") as it pokes above the surface when the Minke comes up for air. Minkes are *fast;* blink and you'll miss them.

PECTORAL FIN VARIATIONS

Amount and shape of white varies in individual whales

Minkes breach more often than larger rorqual whales

UNDERSIDE TAIL PROFILE

UPPERSIDE TAIL PROFILE

Rostrum

Blowhole

Eye

Throat pleats

Pectoral fin

Dorsal fin

Tail stock

Tail fluke

MINKE WHALE

BLUE WHALE

Balaenoptera musculus

Description: A massive blue-gray whale with a proportionately small dorsal fin placed far to rear of its back. Dorsal fin is often not visible until well after whale has begun to dive. The head, when viewed from above, is U-shaped and wide, with a large splashguard ridge just in front of the blowholes. The back is lightly mottled and may show a subtle chevron pattern behind the blowholes. **Red List – Endangered**

Habits: Generally shy; will dive or swim away when approached. The spout is very tall and narrow. A strong swimmer, the Blue Whale can reach almost 20 knots when pressed. When diving, often rolls out its tail flukes a few feet above the water. Typical surface behavior is a blow every 20–30 seconds for several minutes, followed by a shallow dive of five to ten minutes. Favors coastal waters in the Gulf of St. Lawrence and Strait of Belle Isle and, in winter, deep waters near the continental shelf.

Range: Blue Whales can occur anywhere along the East Coast from Cape Hatteras south but are very rarely seen in these waters, and then only in the colder months, well offshore along the continental shelf or in offshore canyons. Known in the Gulf of Mexico only from two strandings in Texas and Louisiana.

Similar species: A large Fin Whale may suggest the Blue Whale, but Blues are so much bigger that they are usually unmistakable when sighted. In diving, the Fin Whale dorsal fin is visible well before the tail stock rolls up, whereas in the Blue Whale, the dorsal is placed so far back that it appears only just before the tail stock submerges.

Size: 30–100 ft. (9.1–30.4 m); adults now average 70–85 ft. (21.3–25.9 m), well under historical maximum sizes.

30–100 ft. (9.1–30.4 m)

BLUE WHALE

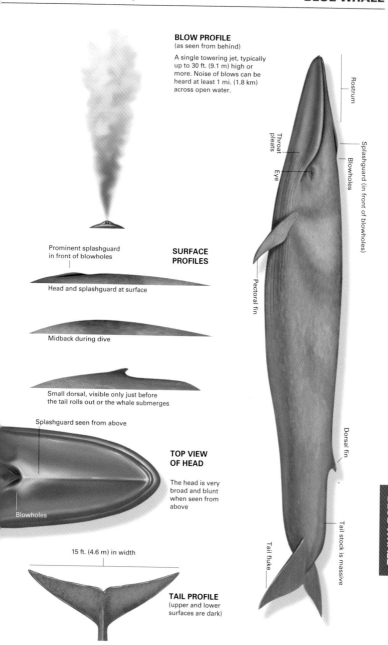

BLOW PROFILE
(as seen from behind)

A single towering jet, typically up to 30 ft. (9.1 m) high or more. Noise of blows can be heard at least 1 mi. (1.8 km) across open water.

Rostrum

Throat pleats

Splashguard (in front of blowholes)

Blowholes

Eye

Pectoral fin

SURFACE PROFILES

Prominent splashguard in front of blowholes

Head and splashguard at surface

Midback during dive

Small dorsal, visible only just before the tail rolls out or the whale submerges

Dorsal fin

Splashguard seen from above

TOP VIEW OF HEAD

The head is very broad and blunt when seen from above

Blowholes

Tail stock is massive

15 ft. (4.6 m) in width

Tail fluke

TAIL PROFILE
(upper and lower surfaces are dark)

BLUE WHALE

Balaenoptera borealis

Description: A spotted blue-gray rorqual smaller than the related Fin and Blue Whales. The back, distinctly lighter than that of the Fin Whale, is typically marked with a sparse pattern of oblong spots. Head and jaws are dark on both sides. A subtle chevron pattern may be visible on the upper back and flanks behind the eyes. Dorsal fin is proportionately large, typically with a tall, falcate shape. Light gray or white belly. **Red List – Endangered**

Habits: Fast and erratic, often changing direction quickly when feeding and swimming in a zigzag pattern. One of the fastest whales, the Sei can reach a speed of 26 knots. A shallow diver, it skims surface waters for food. Rarely arches its back or rolls out its tail when diving. Typically reveals the middle of the back and flanks for longer than in other large rorquals, and when diving seems simply to sink below the surface. Often allows a close approach but then typically speeds away from boats. The Sei's natural history and distribution are poorly known.

Range: Sei Whales are uncommon but regular all along the US Atlantic Coast, drifting northward in the warmer months and into southern waters in the winter months. Seis are most commonly encountered in temperate seas (neither tropical nor polar) at the boundary of the continental shelf where it drops off into much deeper waters or in deep "canyon" areas closer to shore, such as the Baltimore and Norfolk Canyons.

Similar species: Very similar to the Fin Whale and, in southern waters, to Bryde's Whale. The Fin has a white right jaw, is larger, and is darker overall. Bryde's has three ridges on the head, whereas the Sei and Fin have one ridge in front of the splashguard.

Size: 22–50 ft. (6.7–15.2 m); average adult length 35–40 ft. (10.7–12.2 m).

22–50 ft. (6.7–15.2 m)

SEI WHALE

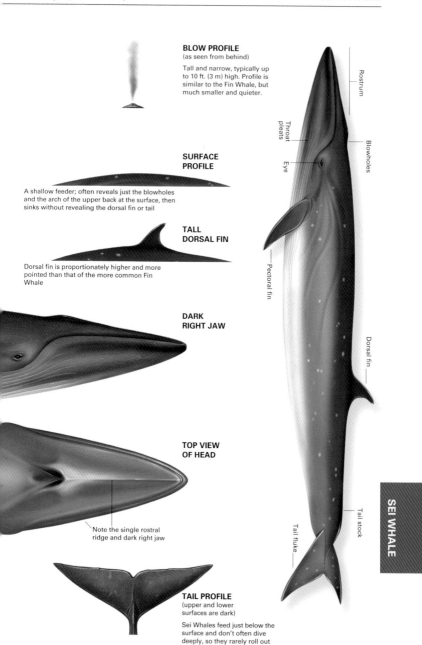

BLOW PROFILE
(as seen from behind)

Tall and narrow, typically up to 10 ft. (3 m) high. Profile is similar to the Fin Whale, but much smaller and quieter.

Rostrum

Throat pleats

Blowholes

Eye

SURFACE PROFILE

A shallow feeder; often reveals just the blowholes and the arch of the upper back at the surface, then sinks without revealing the dorsal fin or tail

Pectoral fin

TALL DORSAL FIN

Dorsal fin is proportionately higher and more pointed than that of the more common Fin Whale

DARK RIGHT JAW

Dorsal fin

TOP VIEW OF HEAD

Note the single rostral ridge and dark right jaw

Tail stock

Tail fluke

SEI WHALE

TAIL PROFILE
(upper and lower surfaces are dark)

Sei Whales feed just below the surface and don't often dive deeply, so they rarely roll out the tail above the water

Description: A blue-gray rorqual with three ridges atop the head forward of the blowholes. Similar to the Sei Whale in size and coloration but restricted to warm waters. The three head ridges originate just forward of the blowholes and are the Bryde's most reliable field marks. **Red List – Data deficient**

Habits: Unlike the surface-skimming Sei Whale, this species is a deep diver and can stay underwater for 20 minutes or more feeding on fish and krill. In areas of cold-current convergence and upwellings, Bryde's Whales will feed in groups. Recent studies have shown a movement from inshore to offshore waters based on seasonal fish movement and plankton blooms. The deep-diving habit is reflected in the Bryde's surface behavior. After a deep dive, its head will often surface at a steep angle, jutting well above the surface. When diving, it bends sharply, exposing much of its back, but it rarely rolls out its tail flukes. Sometimes displays curiosity and may approach boats.

Range: Present along the entire US Atlantic Coast, but unlike the other large whale species, Bryde's Whale is actually more common in warmer waters than in colder northern waters. Bryde's Whale is present in the northern Gulf of Mexico in small numbers (estimated to be 40–50 individuals), particularly in the deeper waters of the DeSoto Canyon area offshore of the Florida Panhandle and the Alabama coast.

Similar species: Look for the three head ridges—Sei and Fin Whales have only one. Very similar to the Sei, but in our area the Sei is most frequent in the North Atlantic in summer, whereas Bryde's Whale rarely moves north of Virginia. The Fin Whale shows a white right jaw and is much larger and darker.

Size: 36–48 ft. (11–14.6 m); average adult length 40 ft. (12.2 m), smaller than the Sei.

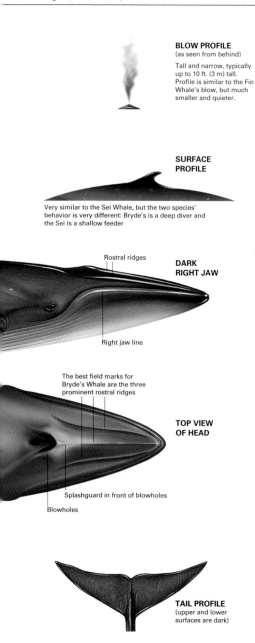

BLOW PROFILE
(as seen from behind)

Tall and narrow, typically up to 10 ft. (3 m) tall. Profile is similar to the Fin Whale's blow, but much smaller and quieter.

Rostrum

Throat pleats

Eye

Blowholes

SURFACE PROFILE

Very similar to the Sei Whale, but the two species' behavior is very different: Bryde's is a deep diver and the Sei is a shallow feeder

Pectoral fin

Rostral ridges

DARK RIGHT JAW

Right jaw line

The best field marks for Bryde's Whale are the three prominent rostral ridges

TOP VIEW OF HEAD

Dorsal fin

Splashguard in front of blowholes

Blowholes

TAIL PROFILE
(upper and lower surfaces are dark)

Tail stock

Tail fluke

BRYDE'S WHALE

329

HUMPBACK WHALE   *Megaptera novaeangliae*

Description: A large blue-black rorqual with very long, mostly white flippers. Variably marked on the head, belly, and tail with white or gray spots or patches. Top of head shows three rows of knobs, making a spy-hopping Humpback look like a giant black pickle. The tail is very wide, with a ragged trailing edge on flukes. **Red List – Least concern**

Habits: Highly inquisitive, readily approaches boats. The Humpback Whale exhibits a wide range of sometimes spectacular social and feeding behaviors at the surface, including breaching, lob-tailing, and flipper-slapping. Humpbacks often feed in groups, where the whales cooperate to surround schools of small fish with "nets" or "clouds" of bubbles blown from their blowholes while underwater (bubble-netting). The blow is low and bushy. The Humpback often rolls its tail high out of the water at the beginning of a dive.

Range: Along the southeastern Atlantic Coast, Humpback Whales are most often seen in migration, moving from northern waters in summer to their winter feeding and calving grounds in the Lesser Antilles. Humpbacks will enter relatively shallow waters within a mile of shore and can often be spotted from the beach in places like the Outer Banks of North Carolina and off the Sea Islands of Georgia. Humpbacks have been seen in the Gulf of Mexico, but only rarely.

Similar species: At a distance, may suggest the Fin or Sperm Whale. The Fin Whale rarely rolls out its tail when diving and does not dive with the sharp "hump" bending at midbody like the Humpback. The Sperm Whale does roll out its tail like the Humpback, but its tail has a quite different shape and rarely shows light patches on the underside. The Sperm Whale has no dorsal fin.

Size: Average adult length 35–45 ft. (10.7–13.7 m).

Tail profiles of the three whale species that frequently "roll out" their tails when diving.

Humpback Whale

Right Whale

Sperm Whale

Typically 35–45 ft. (10.7–13.7 m)   **HUMPBACK WHALE**

BLOW PROFILE
(as seen from behind)

Low and bushy, split left-right.
May appear as a single jet
when seen from the side.
Typically 10 ft. (3 m) high

SURFACE PROFILE

The classic "humped back"
profile, seen as the whale
bends to dive deeply. Note
wide variation in shape and
color of the dorsal fin

**PECTORAL FIN
VARIATIONS**

Mostly white above and
below. Markings are
unique to each whale

Flippers are often raised
vertically above the
surface and slapped
down, apparently a form
of communication

**UNDERSIDE
TAIL PROFILES**

Patterns on the
underside of the tail
range from almost all
white to almost all black.
Each whale's markings
are unique and persist
throughout life, enabling
scientists to identify
individuals.

**UPPERSIDE
TAIL PROFILE**

Labels on whale diagram: Rostrum, Throat pleats, Blowhole, Eye, Pectoral fin, Dorsal fin, Tail stock, Tail fluke

**HUMPBACK
WHALE**

331

Description: Dark gray to black backs and sides, without any dorsal fin or hump. Most individuals in our area also have black bellies. The large white or gray callosities (wartlike structures on the head) are often apparent as the head shows above the water. These outgrowths are inhabited by numerous whale lice that sometimes give the callosities a pink or yellow-orange tint. Very large lower jaw line shows at sides of head when at the surface. Broad flippers with obvious finger structure visible. Uniformly black tail with very smooth, fine-pointed flukes. **Red List – Endangered**

Habits: When feeding, this whale swims slowly at the surface with its mouth open. Often sluggish and surprisingly docile, sometimes suggesting the black upturned hull of a sailboat more than a living whale. However, Rights can also be very energetic and acrobatic at the surface, breaching, rolling, tail-slapping, and spy-hopping. Often rolls its tail high above the surface when diving.

Range: Our small US population of Northern Right Whales feeds in the Gulf of Maine in the summer months and migrates south to waters off the Georgia and Florida Atlantic coasts to winter and give birth to calves. Right Whales are mostly likely to be seen in spring and fall migrations between these two areas. Virtually absent from the Gulf of Mexico, although an individual was spotted off Ft. Walton Beach in Florida in 2004.

Similar species: The tail roll at the beginning of a deeper dive may suggest the Humpback or Sperm Whale. The Humpback Whale has a wide, ragged-looking tail usually marked with white patches. The Sperm Whale has a black tail with a different, more bluntly triangular fluke shape.

Size: Average adult length 35–50 ft. (10.7–15.2 m).

BLOW PROFILE
(as seen from behind)

Thick and bushy, split when seen from behind or in front. Typically 15 ft. (4.6 m) high.

The foremost upper callosity on the head is called the "bonnet"

Side view of head at surface

SURFACE PROFILE

When coming to the surface to breathe will show much of the head and the unique white callosities

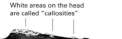

Typical "capsized boat" profile, with a smooth, unmarked black back and no dorsal fin

White areas on the head are called "callosities"

Sometimes brings much of the head above the surface

PECTORAL FIN

When rolling at the surface will often wave pectoral fins in the air before slapping them down onto the surface

A single tail fluke poking up is sometimes the only visible sign of a Right Whale rolling at the surface

TAIL PROFILE
(upper and lower surfaces are black)

Often rolls the tail completely above the surface before beginning a deep dive

"Bonnet" callosity

Eye

Blowhole

Smooth back with no dorsal fin

Pectoral fin

Tail stock

Tail fluke

NORTHERN RIGHT WHALE

333

Physeter macrocephalus

Description: A large, deep gray to black whale with a massive square head and narrow, many-toothed jaw below. No true dorsal fin, but a ridge of irregular bumps along the final third of the back. Pectoral fins are short and stubby. The skin looks wrinkled or corrugated. The blowhole is set at an angle on the left side of the top of the head, giving the blow a distinct leftward and forward projection. **Red List – Vulnerable**

Habits: Most commonly found feeding in the waters above submarine canyons and along the edge of the continental shelf. Often seen in groups moving at a surface speed of 4–5 knots. Swims at the surface in a rhythmic pattern of 30–50 blows before diving and then can be down for as long as an hour. The first exhalation after a long dive is very loud and can be heard over great distances at sea.

Range: Sperm Whales are uncommon but regular all along the US Atlantic Coast, drifting northward in the warmer months and into southern waters in winter. Sperm Whales are most often seen at the boundary of the continental shelf where it drops off into much deeper waters or in deep "canyon" areas closer to shore, such as Virginia's Norfolk Canyon and the DeSoto Canyon off the Florida panhandle. Present in the northern Gulf of Mexico all year, but in smaller numbers, and they are mostly seen in the summer months. Sperm Whales are deep divers and are rarely seen close to shore unless that water is unusually deep.

Similar species: May be confused with the Humpback Whale at long distances, but the square head profile and lack of a dorsal fin are distinctive. The Humpback has long white pectoral fins and usually shows white patches on its tail flukes when diving.

Size: 30–60 ft. (9.1–18.3 m). Sperm Whale populations today rarely reach the large sizes reported by whalers in the past. Adult male averages 50 ft. (15.2 m), female 35 ft. (10.7 m).

30–60 ft. (9.1–18.3 m)

SPERM WHALE

BLOW PROFILE
(as seen from behind)

Low and bushy,
projected forward and
to the left. Typically
7 ft. (2.1 m) high, but
larger males can blow
up to 15 ft. (4.6 m).

SURFACE PROFILE

Side view of Sperm Whale
surfacing. Note the blunt
head profile and forward-
angled blow.

No true dorsal fin but usually a distinct "dorsal hump,"
followed by several smaller humps, or "crenulations"

After breathing at the
surface for 10–20 minutes,
can dive for over an hour.
When preparing for a
deep dive, usually rolls its
tail high into the air as it
leaves the surface at an
almost vertical angle.

TAIL PROFILE
(upper and lower
surfaces are dark)

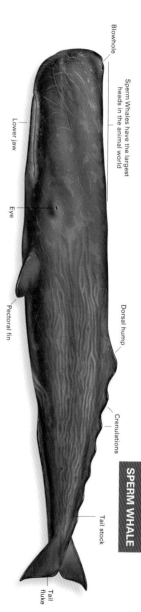

Blowhole

Sperm Whales have the largest heads in the animal world

Lower jaw

Eye

Pectoral fin

Dorsal hump

Crenulations

Tail stock

Tail fluke

SPERM WHALE

LONG-FINNED PILOT WHALE *Globicephala melas*

Description: A medium-sized toothed whale, all black except for dark gray to medium gray "anchor" pattern on midline of belly between pectoral fins. Distinctive dorsal fin, low with very long base, especially noticeable in adult males. Dorsal fin is often hooked in older whales, which may also show lighter "saddle" mark just behind pectoral fin. Pectoral fins have a "wrist" bend, especially in mature individuals. Mature male has a very bulbous forehead profile. Female's head is also blunt but lacks the very rounded profile of the male.

Habits: A very social species. Most often seen in schools of 10–20 individuals, but may also appear in groups of hundreds. Favors cold waters near the continental shelf, well offshore. Though common, the Long-Finned Pilot Whale's offshore habits make it an infrequent sight on coastal whale-watching excursions. Herds of pilot whales regularly strand all along the Atlantic Coast.

Range: The Long-Finned Pilot Whale is the dominant pilot whale north of the Delmarva Peninsula. An area of overlap exists off Maryland to North Carolina where both Long-Finned and Short-Finned Pilot Whales are possible, but south of New Jersey most pilot whales you encounter will be Short-Finned.

Similar species: Separated from the Short-Finned Pilot Whale only by geographic range; the Short-Finned is rarely seen north of New Jersey. Pilot whales seen north of Cape Hatteras are most likely to be Long-Finned.

Size: 10–20 ft. (3–6.1 m); adult male averages 17–19 ft. (5.2–5.8 m), female 13 ft. (4 m).

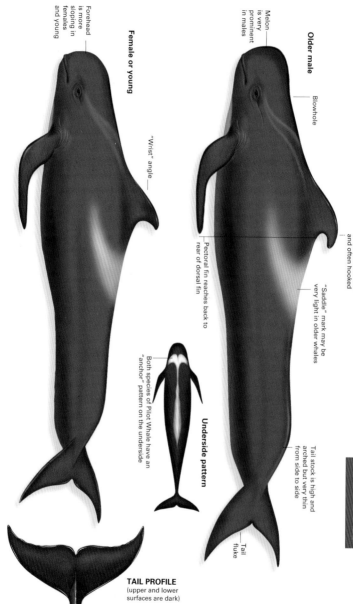

Older male

Melon is very prominent in males

Blowhole

Dorsal fin is broad and often hooked

"Saddle" mark may be very light in older whales

Pectoral fin reaches back to rear of dorsal fin

Tail stock is high and arched but very thin from side to side

Tail fluke

Female or young

Forehead is more sloping in females and young

"Wrist" angle

Underside pattern

Both species of Pilot Whale have an "anchor" pattern on the underside

TAIL PROFILE
(upper and lower surfaces are dark)

Description: A near duplicate of the Long-finned Pilot Whale. Range is the key factor in field identification. In good light, look for a gray "saddle" mark behind the dorsal fin and more white in throat area than in the Long-Finned. Pectoral fins are shorter in proportion to body size, and dorsal fin is typically higher and more falcate than in the Long-Finned. The male's large, bulbous head grows with age. **Red List – Data deficient**

Habits: Most commonly seen in pods of 10–30 individuals, but some pods can be much larger. Mainly an offshore whale, occasionally coming near shore. Large groups are regularly stranded on the Atlantic Coast for unknown reasons. Thought to feed on squid in deep offshore waters and near-shore canyon areas, mainly at night.

Range: The Short-Finned Pilot Whale is the dominant pilot whale south of Cape Hatteras, but an area of overlap exists off Virginia and North Carolina where both Long-Finned and Short-Finned Pilot Whales are present. Rare north of Delaware. The only pilot whale species normally seen in the Gulf of Mexico.

Similar species: Can be separated from the Long-Finned Pilot Whale only by geographic range; the Long-Finned is rarely seen south of New Jersey. All pilot whales south of Maryland are almost certainly Short-Finned.

Size: 9–19 ft. (2.7–5.5 m); adult male averages 17–19 ft. (5.2–5.8 m), female 13 ft. (4 m).

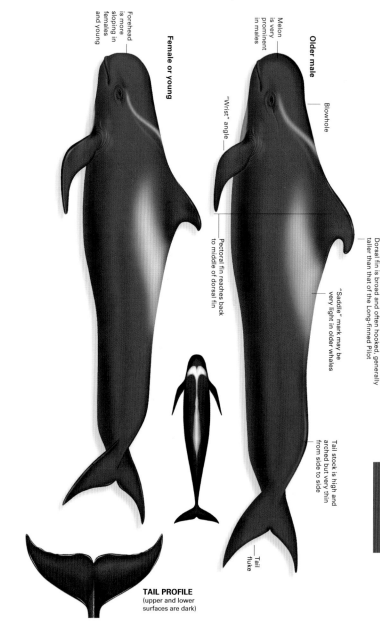

Female or young

Forehead is more sloping in females and young

Melon is very prominent in males

Older male

"Wrist" angle

Blowhole

Pectoral fin reaches back to middle of dorsal fin

Dorsal fin is broad and often hooked, generally taller than that of the Long-finned Pilot

"Saddle" mark may be very light in older whales

Tail stock is high and arched but very thin from side to side

Tail fluke

TAIL PROFILE
(upper and lower surfaces are dark)

Description: A very large black dolphin. Relatively small tapered head, sloping forehead profile, and "wrist" bends in pectoral fins are the best field marks for this species. Mostly black except for a small patch of gray on chest between pectoral fins. Large teeth are often visible at close ranges.

Habits: Favors deep ocean waters and is rarely seen north of Maryland. The only "blackfish" species that rides bow waves. False Killer Whales are never common in our area, but they appear sporadically in large groups that are very energetic and likely to be noticed if present. They breach, lob-tail, and spy-hop at the surface and are curious about boats. False Killers are reported to attack other dolphin species and sometimes young Humpback Whales.

Range: Common off the coast from Cape Hatteras to the Caribbean as well as in the Gulf of Mexico, but only in deep waters well offshore (3,300 ft., or 1,000 m) and along the edge of the continental shelf.

Similar species: Slimmer and smaller-headed than the Killer Whale (Orca) and lacks any trace of white. Darker than all other toothed whale species except pilot whales. Pilot whales have large, bulbous heads and lack the tall, falcate dorsal fin of the False Killer Whale.

Size: 9–19 ft. (2.7–5.8 m). Very large, comparable in size to the Killer Whale and pilot whales. Average adult length 15 ft. (4.6 m).

FALSE KILLER WHALE

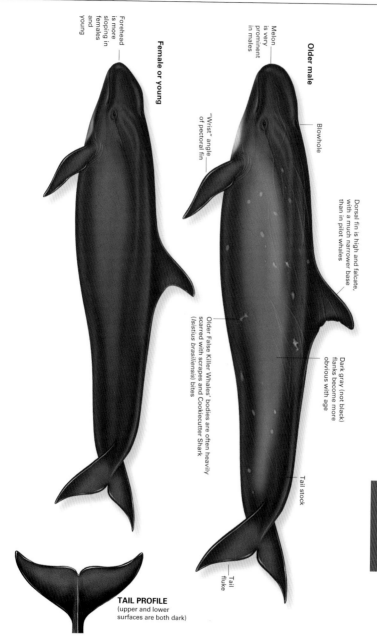

Older male

Melon is very prominent in males

Blowhole

"Wrist" angle of pectoral fin

Dorsal fin is high and falcate, with a much narrower base than in pilot whales

Older False Killer Whales' bodies are often heavily scarred with scrapes and Cookiecutter Shark (*Isistius brasiliensis*) bites

Dark gray (not black) flanks become more obvious with age

Tail stock

Tail fluke

Female or young

Forehead is more sloping in females and young

TAIL PROFILE
(upper and lower surfaces are both dark)

SEPARATING THE "BLACKFISH"

"Blackfish" is the informal term used to describe these large black dolphin species. Our four blackfish all favor offshore waters, although they may sometimes be seen near shore. We have grouped them here for easy comparison. At sea it can be quite difficult to tell one blackfish species from another.

KILLER WHALE (ORCA) *Orcinus orca*

Uncommon in this area. The white side patches and massive girth of the Killer Whale are the best field marks. Remember that females and younger Orcas of both sexes lack the huge dorsal fin of the adult male. Female dorsal fins are similar to the False Killer Whale's. The low, bushy blow is sometimes visible in cold air. **Red List – Data deficient**

FALSE KILLER WHALE *Pseudorca crassidens*

All black; never shows any trace of white. Much smaller and trimmer than the Killer Whale (Orca). Note the narrow pectoral fins, with distinct "wrist" angles and sharp pointed ends. Much more common than the Killer Whale, especially in waters south of Virginia.

LONG-FINNED PILOT WHALE *Globicephala melas*

Rare south of Cape Hatteras. The low, broad-based dorsal fin and bulbous forehead profile are the best field marks for this species. At sea it is almost impossible to separate the Long-Finned Pilot Whale from the Short-Finned Pilot Whale by sight. Note the "saddle" mark behind the dorsal fin, which grows lighter with age.

SHORT-FINNED PILOT WHALE *Globicephala macrorhynchus*

The proportionately short pectoral fins are the only useful field mark. The Short-Finned Pilot Whale has a more southern distribution, and *most* pilot whales south of New Jersey can be identified as Short-Finneds. North of Long Island, the species will almost certainly be the Long-Finned Pilot Whale. **Red List – Data deficient**

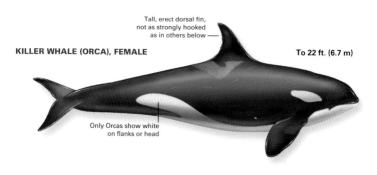

KILLER WHALE (ORCA), FEMALE

Tall, erect dorsal fin, not as strongly hooked as in others below —

To 22 ft. (6.7 m)

Only Orcas show white on flanks or head

FALSE KILLER WHALE

Dorsal fin is typically smaller and more hooked than in Orcas

To 19 ft. (5.8 m)

Narrow pectoral fins; flanks are solid black

"Wrist" angle

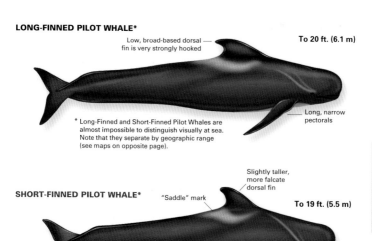

LONG-FINNED PILOT WHALE*

Low, broad-based dorsal fin is very strongly hooked

To 20 ft. (6.1 m)

Long, narrow pectorals

* Long-Finned and Short-Finned Pilot Whales are almost impossible to distinguish visually at sea. Note that they separate by geographic range (see maps on opposite page).

SHORT-FINNED PILOT WHALE*

Slightly taller, more falcate dorsal fin

"Saddle" mark

To 19 ft. (5.5 m)

Shorter pectorals

White patch is more visible

"BLACKFISH"

GRAMPUS (RISSO'S DOLPHIN) *Grampus griseus*

Description: Also called Risso's Dolphin. A large-headed, blunt-faced dolphin reminiscent of pilot whales in general shape. Forehead is blunt and sloping, without a bulbous "melon" shape. Light gray body with dark fins. Perhaps the most diagnostic feature is the scarring that occurs all over the back and sides, apparently created by other Grampus while fighting. The network of scars on older adults makes their backs very light gray. **Red List – Least concern**

Habits: Grampus favor deep warm water well offshore, such as the Gulf Stream off Cape Hatteras. They are very active at the surface, where they breach, lob-tail, and spy-hop, often in large pods of 30 or more individuals. Grampus feed mainly on squid, and their underparts often show sucker marks.

Range: Common off the coast from Cape Hatteras to the Caribbean and in the Gulf of Mexico, but only in deep waters well offshore (3,300 ft, or 1000 m) and along the edge of the continental shelf.

Similar species: Schools of pilot whales are often encountered in the same areas frequented by the Grampus. Pilot whales are much darker, have much more bulbous heads, and have dorsal fins that are much lower and broad-based than the Grampus. May be separated from other dolphin species by its overall coloration and blunt head shape.

Size: 8–13 ft. (2.5–4 m); average adult length 10 ft. (3 m) and weight 650 lbs. (295 kg).

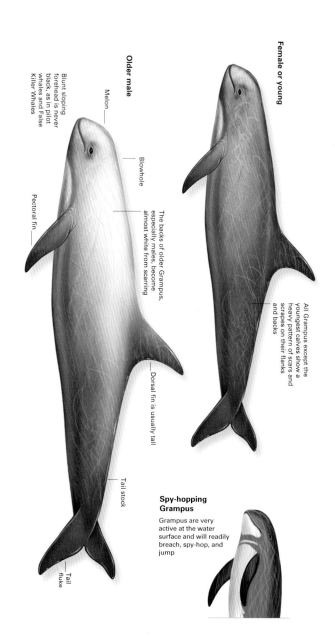

Female or young

Older male

Melon

Blowhole

Pectoral fin

Blunt sloping
forehead is never
black, as in pilot
whales and False
Killer Whales

The backs of older Grampus,
especially males, become
almost white from scarring

Dorsal fin is usually tall

All Grampus except the
youngest calves show a
heavy pattern of scars and
scrapes on their flanks
and backs

Tail stock

Tail
fluke

**Spy-hopping
Grampus**

Grampus are very
active at the water
surface and will readily
breach, spy-hop, and
jump

GRAMPUS

345

Description: Unmistakable at all but the farthest ranges. Deep black back is divided from almost pure white belly in a hard line, as is white patch behind eye. Dorsal fin of male may reach 6 ft. (1.8 m) in height and is often the first noticeable field mark at a distance. Body is much more massive than any other dolphin species. **Red List – Data deficient**

Habits: An unusual but regular visitor to the North Atlantic Coast. Rare south of Long Island. Western Atlantic Killer Whales are much less well known than their Pacific Ocean counterparts. They are thought to follow Bluefin Tuna populations along the Atlantic Coast, and they appear in the Gulf of Maine at about the same time as the Bluefins (mid-July). Killer Whales are known to attack Humpback Whales, whose tails often show Orca toothmarks.

Range: Uncommon to rare off the entire Atlantic Coast and in the Gulf of Mexico. Killer Whales are one of the most widely distributed whale species, so they can and do appear almost anywhere, but only rarely in our area and in very small numbers. The National Marine Fisheries Service estimates that there may be about 150 Killer Whales in the entire Gulf.

Similar species: The female Orca has a smaller dorsal fin similar in shape to that of the False Killer Whale and to many dolphin species. The female Orca is always much longer and very much heavier than dolphins, and no dolphin in our area shows such bold white patches on the head and flanks. Pilot whales have no white patches, and their dorsal fins are low, often hooked over at the tip, and very broad at the base.

Size: 12–30 ft. (3.7–9.1 m); adult male averages 19–22 ft. (5.8–6.7 m), with a massive girth; female is considerably smaller and slimmer, averaging 16 ft. (4.9 m).

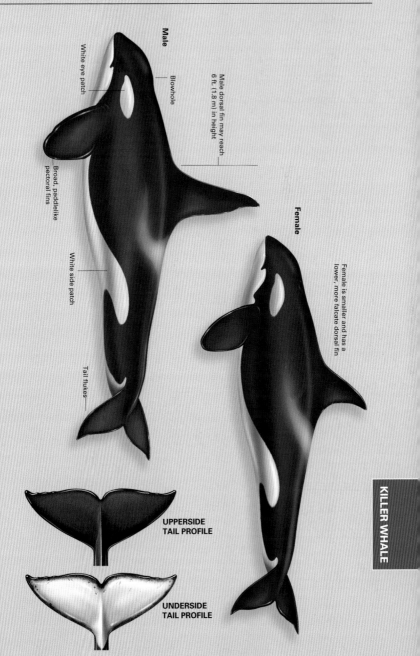

Male

White eye patch

Blowhole

Male dorsal fin may reach
6 ft. (1.8 m) in height

Broad, paddlelike
pectoral fins

White side patch

Tail flukes

Female

Female is smaller and has a
lower, more falcate dorsal fin

**UPPERSIDE
TAIL PROFILE**

**UNDERSIDE
TAIL PROFILE**

KILLER WHALE

347

SPERM WHALES

The behavior and natural history of these two closely related whales are not well understood. Both species may be more common than the sparse record of sightings suggests, because both are often stranded along the southern Atlantic Coast. The Pygmy is known from northern waters mostly by strandings; it is rarely seen north of Cape Cod.

PYGMY SPERM WHALE
Kogia breviceps

A very small whale, blunt-headed, with uniform gray coloration on back and sides and lighter gray belly. Very small dorsal fin. Small underslung jaw, as in the Sperm Whale. Only the lower jaw contains teeth. Note "false gill" marks behind and below eye. Inconspicuous; rarely approaches boats. May lie dormant at the surface and allow boats to approach closely. When startled, emits a cloud of reddish fluid from the anus and dives quickly. Sometimes leaps out of the water. Often seen alone or in small groups of up to 12 individuals. Sightings are uncommon, but the Pygmy is regularly spotted off the coast of North Carolina. Thought to prefer deep warm waters well offshore. **Size:** 9–11 ft. (2.7–3.4 m).

DWARF SPERM WHALE
Kogia simus

The natural history, range, and habits of the Dwarf Sperm Whale are poorly understood. Like the Pygmy Sperm Whale, the Dwarf favors warm waters well offshore, and the two species are almost impossible to distinguish in the field. The Dwarf has a much larger, more falcate dorsal fin. When these two species are stranded, the dorsal fin and "false gill" marks give them a sharklike appearance. Some observers have speculated that the gill marks may have evolved to make these two whales look even more like sharks and thus discourage predators. Like the Pygmy, the Dwarf Sperm Whale discharges reddish fluid from the anus when startled. **Size:** 7–8 ft. (2.1–2.4 m).

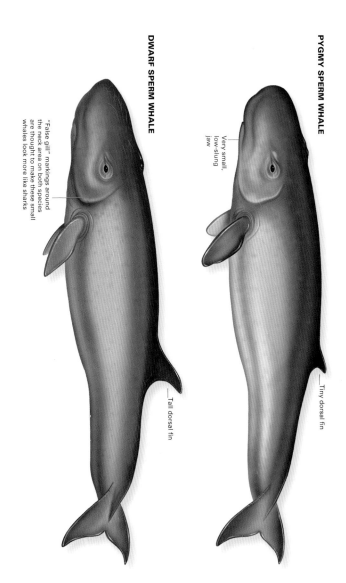

PYGMY SPERM WHALE

Very small, low-slung jaw

Tiny dorsal fin

DWARF SPERM WHALE

"False gill" markings around the neck area on both species are thought to make these small whales look more like sharks

Tall dorsal fin

BEAKED WHALES

The medium-sized whales in this fairly large genus are rarely seen and are poorly understood. These species are deep-ocean residents and are inconspicuous at the surface, so they are often overlooked.

In true beaked whales, the male possesses an odd, enlarged tooth ("beak") that appears a fourth to halfway along the jaw and sticks out beside the upper jaw. The occurrence of beaked whales is unpredictable. These whales are so shy that the sound of an oncoming boat will cause them to flee. At sea they are inconspicuous and very difficult to identify with precision.

BLAINVILLE'S BEAKED WHALE *Mesoplodon densirostris*

Sometimes called the Dense-Beaked Whale. Male has a large, arched jawline with a tooth protruding from its ridge at the halfway point. Elongated rostrum. Flippers far to the front. Probably common but very shy of boats. Mainly a warm-water species, but individuals wander north with the Gulf Stream. Usually travels and feeds in small groups. **Size:** Average adult length 15–20 ft. (4.6–6.1 m). **Red List – Data deficient**

TRUE'S BEAKED WHALE *Mesoplodon mirus*

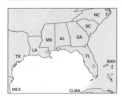

Small head, short beak, and bulging forehead. Flippers far to the front. Unlike other beaked whales, tail fluke has central notch. Seen regularly off Carolinas, but lack of sightings elsewhere may be due to the shyness, not rarity. Feeds mainly on squid. **Size:** Average adult length 16–18 ft. (4.9–5.5 m). **Red List – Data deficient**

SOWERBY'S BEAKED WHALE *Mesoplodon bidens*

An unusual northern species that barely enters our range, being more common north of Cape Hatteras. Gray to dark olive back normally shows many scars. Falcate dorsal fin variable in shape and size. Most easily separated from other beaked whale species by its long beak. **Size:** Average adult length 15–20 ft. (4.6–6.1 m). **Red List – Data deficient**

15–20 ft. (4.6–6.1 m)

BEAKED WHALES

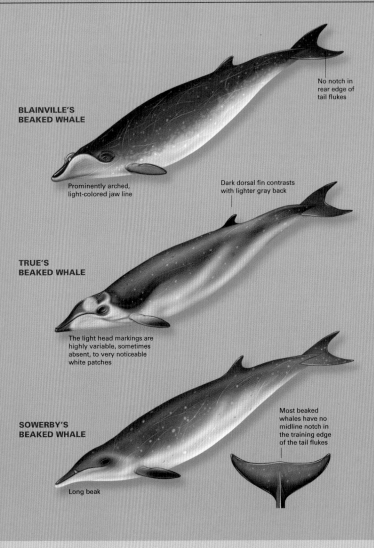

**BLAINVILLE'S
BEAKED WHALE**

No notch in
rear edge of
tail flukes

Prominently arched,
light-colored jaw line

Dark dorsal fin contrasts
with lighter gray back

**TRUE'S
BEAKED WHALE**

The light head markings are
highly variable, sometimes
absent, to very noticeable
white patches

**SOWERBY'S
BEAKED WHALE**

Most beaked
whales have no
midline notch in
the training edge
of the tail flukes

Long beak

**BEAKED
WHALES**

COOKIECUTTER SHARK *Isistius brasiliensis*

2 ft. (61 cm)

In warmer waters, the tiny Cookiecutter Shark plagues large fish and whales
by taking small circular bites from their sides, leaving the round, light scars so
commonly seen on many whale species

CUVIER'S BEAKED WHALE
Ziphius cavirostris

Description: Formerly known as the Goosebeaked Whale. The most common beaked whale, especially south of Long Island Sound. White forehead and back in front of dorsal fin. Dorsal fin small, falcate, variable in shape and size, and located behind the midpoint of the back. The snout is short and seems to blend into the head. Color varies from brown or tan to slate gray. Only males have two teeth at tip of the lower jaw. Cuvier's Beaked Whales will roll out their tail flukes before beginning a deep dive. **Red List – Least concern**

Habits: Often seen in groups of 10–25 individuals, primarily in offshore waters. Frequent strandings, especially along the southeastern Atlantic Coast, may indicate that Cuvier's approaches shallow water more often than other beaked whales or could be a direct measure of this whale's abundance. Apparently a deep diver; can stay down from 20 to 40 minutes. Feeds primarily on squid.

Range: Found throughout temperate and tropical waters worldwide, including the Gulf of Mexico. Normally seen in very deep offshore waters near the edges of the continental shelf, but does sometimes wander into waters closer to shore.

Similar species: Minke Whales are similar in size and surface behavior but are much darker and show a dark, pointed head profile quite unlike that of the beaked whales. No other beaked whales show the light or white forehead seen in adult males.

Size: Much larger than dolphins and most other beaked whales; average adult length 18–24 ft. (5.5–7.3 m).

18–24 ft. (5.5–7.3 m)

CUVIER'S BEAKED WHALE

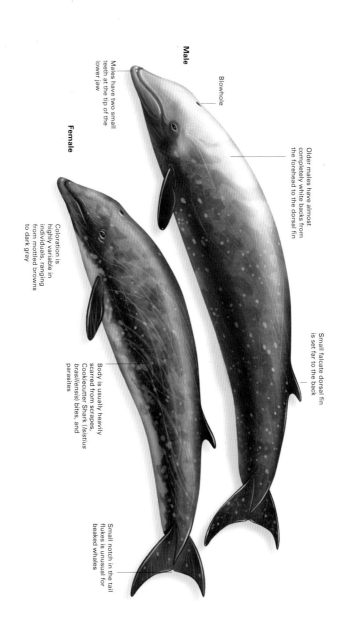

Male

Blowhole

Males have two small teeth at the tip of the lower jaw

Older males have almost completely white backs from the forehead to the dorsal fin

Small falcate dorsal fin is set far to the back

Female

Coloration is highly variable in individuals, ranging from mottled browns to dark gray

Body is usually heavily scarred from scrapes, Cookiecutter Shark (*Isistius brasiliensis*) bites, and parasites

Small notch in the tail flukes is unusual for beaked whales

CUVIER'S
BEAKED WHALE

ATLANTIC WHITE-SIDED DOLPHIN
Lagenorhynchus acutus

A dark, medium-sized dolphin with bold black, white, and mustard-yellow flank markings. Unmistakable at close range. Note that the beak is always dark and that the area behind the dorsal fin is dark with a bright mustard-yellow patch. Unlike the smudgy, indistinct flank patches of the closely related White-Beaked Dolphin (*Lagenorhynchus albirostris*), flank markings are sharply delineated and more contrasting. Common in the Gulf of Maine in deeper waters (150 ft. [48 m] or more), rarely venturing near shore except in strandings. Rarely rides bows, but very active at the surface, often jumping and tail-lobbing. Usually spotted near feeding groups of whales and seabirds. Common from Labrador to Cape Cod. Spotty occurrences south of Cape Cod to Cape Hatteras. **Size:** Average male length 7–9 ft. (2.1–2.7 m), female 6–7 ft. (1.8–2.1 m). **Red List – Least concern**

HARBOR PORPOISE
Phocoena phocoena

The smallest toothed cetacean in our area, with an average length of 5 ft. (1.5 m) or less and typically weighing only about 100 lbs. (45 kg). Often seen in small social groups of 10–15 individuals. Dark above and pale below, chunky, with small pectoral and dorsal fins. Common, especially inshore and north of Cape Hatteras, but not often noticed because of its small size, quiet habits, and shyness around boats. Present in small numbers south of Cape Hatteras, but its small size and inconspicuous surface activity make it hard to spot, especially in choppy waters, where it can easily disappear into wave troughs. **Size:** To 4–6 ft. (1.2–1.8 m). **Red List – Least concern**

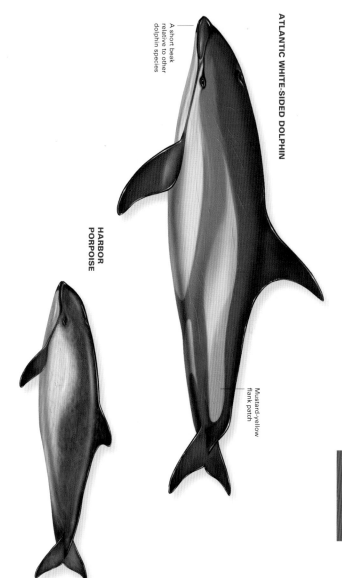

ATLANTIC WHITE-SIDED DOLPHIN

A short beak relative to other dolphin species

HARBOR PORPOISE

Mustard-yellow flank patch

DOLPHINS

355

Description: These two very similar species of *Stenella* dolphins are difficult to distinguish at sea. Young spotted dolphins of both species are unmarked and may be confused with the similar spinner dolphins. Mature spotted dolphins show a distinct spotted pattern on the flanks and back that increases in contrast and complexity with age. Older spotted dolphins often have bellies that are almost as darkly spotted as their backs and backs so spotted that they are almost light gray. When seen underwater from a boat, as when the dolphins are bow-riding, spotted dolphins often show a slight purple or violet coloration.

Habits: These are true pelagic dolphins that favor deep tropical waters well offshore. Both species are fast, high-energy swimmers that often leap high out of the water and readily ride the bow waves of ships. They are thought to feed mainly on Yellowfin Tuna and squid but also take small schooling fish such as herring and anchovies. As in other *Stenella* dolphins, the spotted species are highly social and may appear in groups ranging from a dozen individuals up to herds of several thousand dolphins of various species (usually Spinner, Striped, and Bottlenose Dolphins).

ATLANTIC SPOTTED DOLPHIN   *Stenella frontalis*

Coloration is often lighter and less heavily spotted than the Pantropical Spotted Dolphin, but this is highly variable. Note the much lighter eye stripe. This species is unique to the tropical waters of the Atlantic Ocean. Very active at the surface, often leaping, lob-tailing, and bow-riding. Occurs from Long Island south but is never common in our area. **Size:** To 6–8 ft. (1.8–2.4 m). **Red List – Data deficient**

PANTROPICAL SPOTTED DOLPHIN   *Stenella attenuata*

Back coloration is often darker than the Atlantic Spotted Dolphin, but only in younger individuals. Older Pantropicals are so heavily spotted that the dark back–light belly pattern gives way to a uniformly gray spotted pattern all over the body. Note the heavier eye stripe behind the eye. **Size:** To 6–8 ft. (1.8–2.4 m). **Red List – Least concern**

ATLANTIC SPOTTED DOLPHIN

PANTROPICAL SPOTTED DOLPHIN

SPOTTED DOLPHINS

SPINNER AND STRIPED DOLPHINS

Stenella spp.

These are the most acrobatic of all dolphin species, often leaping high above the surface and "spinning" with their characteristic spiral-rolling leaps. These dolphins prefer warm offshore waters and are often associated with other tropical species like the Pantropical and Atlantic Spotted Dolphins. All are thought to follow schools of Yellowfin Tuna.

SPINNER DOLPHIN

Stenella longirostris

Dorsal fin triangular and very erect, not hooked or falcate. Long, thin beak. Flanks show three distinct color bands: black back, gray sides, and light-colored or white belly. Dark eye stripe runs from eye back to pectoral fins. Lower jaw is uniformly dark. Otherwise very difficult to separate from the Clymene Dolphin. **Size:** To 4.5–7 ft. (1.4–2.1 m). **Red List – Data deficient**

CLYMENE DOLPHIN

Stenella clymene

Once also known as the Short-Snouted Spinner Dolphin. Poorly known, probably because it is so hard to distinguish from the Spinner Dolphin. Dorsal fin shows typical falcate dolphin shape. Beak is shorter and stouter than that of the Spinner. Lower lips and tip of lower jaw are black; rest of lower jaw typically is white. Otherwise nearly identical to the Spinner Dolphin. **Size:** To 4.5–7 ft. (1.4–2.1 m). **Red List – Data deficient**

STRIPED DOLPHIN

Stenella coeruleoalba

A deep-water species normally seen only far out at sea in warm Gulf Stream waters. There are occasional reports from Georges Bank and the southern Gulf of Maine. Look for the bold "bilge line" that runs from the eye to the anus area and the gray flank markings. Very active at surface, porpoising, spinning, breaching, and bow-riding. **Size:** To 4.5–7 ft. (1.4–2.1 m). **Red List – Least concern**

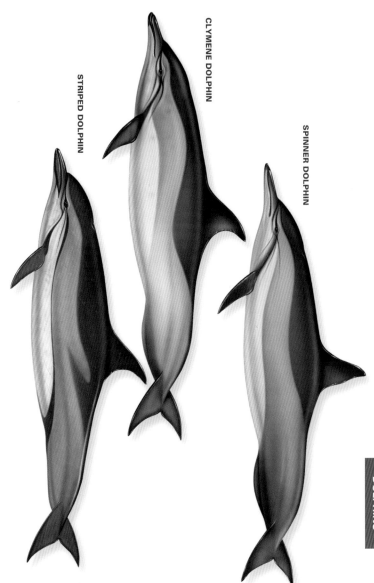

CLYMENE DOLPHIN

STRIPED DOLPHIN

SPINNER DOLPHIN

DOLPHINS

COMMON DOLPHIN
Delphinus delphis

A beautifully patterned animal. Deep gray back with yellowish sides and white underbelly contrasting into gray flanks. No other dolphin has as complex a side pattern. Note also dark rings around eyes, dark line forward of eyes, and dark beak. Found in large schools usually churning up the water as they feed or seemingly "play." They frequently jump clear of the surface and readily ride the bow waves of boats. Locally common in warmer waters throughout the world, these dolphins can appear anywhere off the Atlantic Coast in just about any water depth. However, the largest groups we have witnessed have been over the upwellings around submarine ridges and in the Gulf Stream, well offshore in deep warm waters. In northern waters may be confused with heavier-bodied Atlantic White-Sided Dolphins or White-Beaked Dolphins (*Lagenorhynchus albirostris*). In southern waters may be confused with spotted dolphins or very energetic schools of Bottlenose Dolphins. **Size:** To 6–8 ft. (1.8–2.4 m). **Red List – Least concern**

ROUGH-TOOTHED DOLPHIN
Steno bredanensis

An uncommon warm-water species, rarely encountered north of Cape Hatteras except in Gulf Stream waters far offshore. Uniformly dark back, upper flanks, and tail stock. Back color ranges from dark gray-brown to almost black. Lower sides and belly marked with mottling and spots of lighter gray to white. Smudgy eyeline extends back from eye to base of pectoral fin. Low, sloping forehead merges smoothly into beak without crease at base of beak characteristic of most dolphin species. Teeth are marked with fine vertical grooves that give the species its name. **Size:** To 6–8 ft. (1.8–2.4 m). **Red List – Least concern**

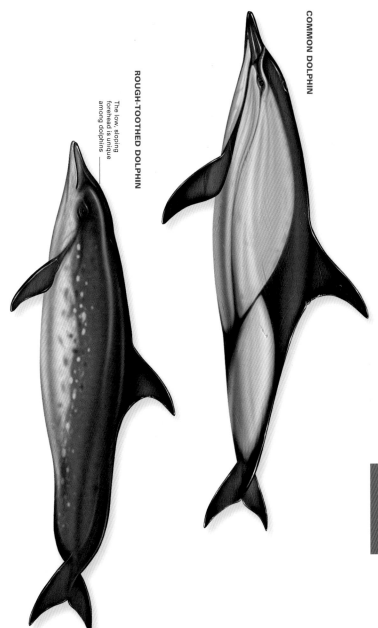

COMMON DOLPHIN

ROUGH-TOOTHED DOLPHIN

The low, sloping forehead is unique among dolphins

BOTTLENOSE DOLPHIN   *Tursiops truncatus*

Perhaps the best known of all cetaceans, based on its television and film exposure. The only large dolphin species that regularly comes near the shore. Runs along shallow coastal waters, readily approaching boats, and enters large estuaries, where it can be viewed from bridges. Neutral to dark gray on the back and sides, grading to a pink-white underbelly. This species occurs in two populations that are closely related but genetically distinct: a larger, darker *offshore* ecotype and a smaller, lighter-colored *coastal* form.

Habits: The Bottlenose is highly social. It often "porpoises" when moving quickly, rolling and jumping free of the ocean surface, and it frequently slaps its tail on the surface (a social signal among dolphins) before diving.

Range: Common all along the US Atlantic Coast, and particularly from Virginia south to the Florida Keys, Bahamas, Caribbean, and throughout the Gulf of Mexico. Found in all waters, from just off the beach to the edge of the continental shelf and beyond.

Size: A large dolphin, with an average weight of 400 lbs in the offshore form. (181 kg). Offshore form averages 7.5–9 ft. (2.3–2.7 m) in length; coastal form is smaller, at 5–7.5 ft. (1.5–2.3 m). **Red List – Least concern**

Bottlenose Dolphins are a common sight close to shore along the Outer Banks, Sea Islands, and Florida Atlantic and Gulf Coasts. The dolphin shown below is typical of the *coastal ecotype* of the Bottlenose Dolphin, which is smaller and lighter-toned than the darker, larger *offshore ecotype*.

P. LYNCH

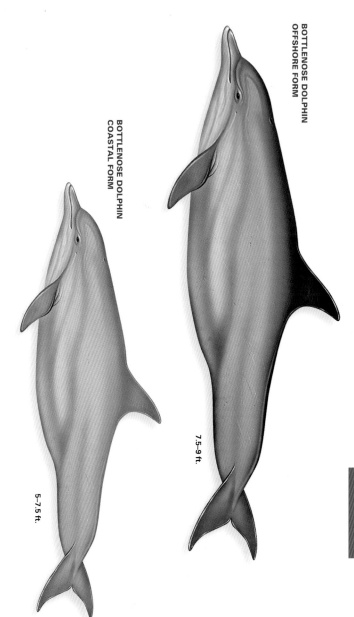

BOTTLENOSE DOLPHIN
OFFSHORE FORM

BOTTLENOSE DOLPHIN
COASTAL FORM

7.5-9 ft.

5-7.5 ft.

363

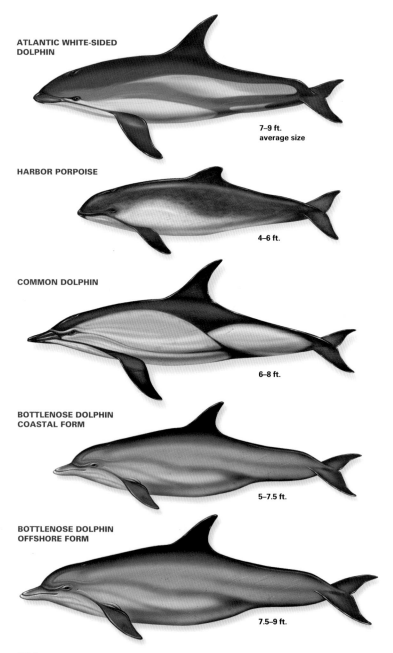

ATLANTIC WHITE-SIDED DOLPHIN

7–9 ft.
average size

HARBOR PORPOISE

4–6 ft.

COMMON DOLPHIN

6–8 ft.

BOTTLENOSE DOLPHIN COASTAL FORM

5–7.5 ft.

BOTTLENOSE DOLPHIN OFFSHORE FORM

7.5–9 ft.

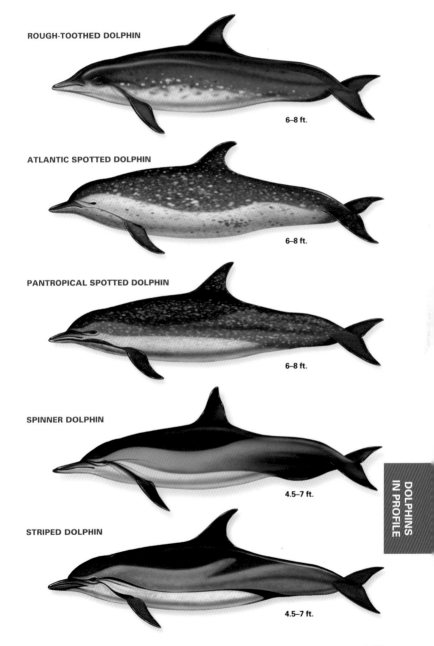

ROUGH-TOOTHED DOLPHIN

6–8 ft.

ATLANTIC SPOTTED DOLPHIN

6–8 ft.

PANTROPICAL SPOTTED DOLPHIN

6–8 ft.

SPINNER DOLPHIN

4.5–7 ft.

STRIPED DOLPHIN

4.5–7 ft.

DOLPHINS IN PROFILE

HARBOR SEAL

Phoca vitulina

Description: The most common and familiar seal along the Atlantic Coast. Overall pelage (fur) color may range from light gray to almost black, with wide variations in the amount and pattern of spotting. Note that the color changes as the pelage dries: wet seals look much darker and grayer than dry seals. Head silhouette is very rounded, with a very short, doglike snout. Forehead slopes into the snout in a concave curve, unlike the heavy convex forehead of the similar Gray Seal.

Habits: Common in inshore waters from the Gulf of St. Lawrence south to Long Island Sound; much less common south of New York. Favors rocky breakwaters, small islands, and jetties as hauling-out points. Very playful and active at the surface. Sometimes shy, but may also be boldly curious and will readily approach boats. Watch for Harbor Seals' small, round heads spy-hopping at the surface.

Range: The primary range for Harbor Seals is the Canadian Maritimes south to Long Island. However, young Harbor Seals wander widely, and the Harbor Seal is regularly spotted in winter along the southeastern coast to the Carolinas. The number of sightings increases every year, perhaps because the whole population is expanding southward.

Similar species: The Gray Seal has a longer, heavier, horselike snout, whereas the short, rounded snout and concave forehead of the Harbor Seal suggest a spaniel puppy. The Gray Seal is also much larger and heavier-bodied.

Size: 3–6 ft. (0.9–1.8 m). Average length 3–5 ft. (0.9–1.5 m) and weight 150–175 lbs. (68–79 kg). Female is about 10 percent smaller than male.

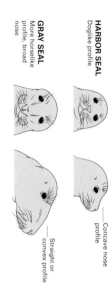

GRAY SEAL
More horselike profile, broad nose

HARBOR SEAL
Doglike profile

——— Concave nose profile

——— Straight or convex profile

Adult

Pup

GRAY SEAL

Halichoerus grypus

Description: A large gray to black seal with a heavy, horselike snout and a convex forehead profile. Pelage (fur) color ranges from light gray to brown or almost black, with wide variations of spot patterns within the population. Pups are typically white to pale gray with very light spots, darkening as they mature.

Habits: Favors colder waters than the Harbor Seal. Gray Seal populations are centered around the Gulf of St. Lawrence and are much less common south of Massachusetts. In recent years the range of the Gray Seal has been expanding southward. Prefers isolated rocky offshore islands, breakwaters, and jetties as hauling-out spots. This preference for isolated offshore haul-out spots makes sightings of the Gray Seal much less frequent than the less shy Harbor Seal.

Range: The Gray Seal is a cold-water species with a normal range well north of the area covered by this guide. However, individual Gray Seals are rarely but persistently seen alive or found dead in the Outer Banks of North Carolina in winter, so there is always a *slim* possibility that a seal you see in North Carolina in winter may be a Gray Seal.

Similar species: A good look at the head shape is crucial to separating our two common seal species. The Harbor Seal has a very short, doglike snout and is much smaller and lighter in weight than the Gray Seal. The pelage patterns of Gray and Harbor Seals are very similar and cannot be used to separate them. Gray Seals are typically much less curious about boats and humans than are the more gregarious Harbor Seals.

Size: 5–8 ft. (1.5–2.4 m). Adult male averages 7 ft. (2.1 m) and 800 lbs. (360 kg). Adult female averages 5–6 ft. (1.5–1.8 m) and 400 lbs. (180 kg).

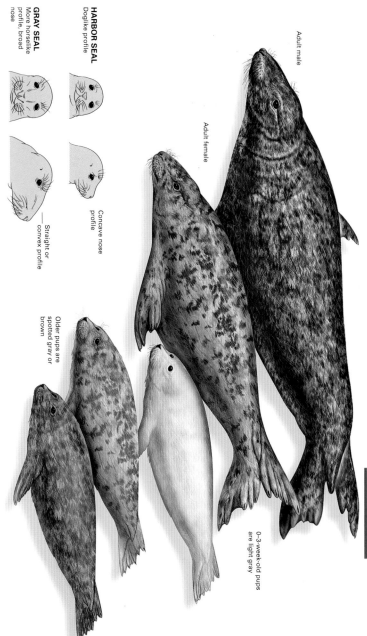

Adult male

Adult female

HARBOR SEAL
Doglike profile

GRAY SEAL
More horselike
profile, broad
nose

Concave nose
profile

Straight or
convex profile

Older pups are
spotted gray or
brown

0–3-week-old pups
are light gray

GRAY SEAL

WEST INDIAN MANATEE

Trichechus manatus

Manatees are large, slow-moving denizens of coastal and riverine areas. In the United States they have become one of the most recognized of all marine mammals, and manatee habitat destruction and motorboat injuries has led to the establishment of protected manatee sanctuaries.

Description: A massive gray-brown mammal with a bulky body that tapers to a flattened, paddle-shaped tail. The back may show the deep scars of boat propellers. The large head has a broad, whiskered snout, and tiny eyes.

Habits: The manatee eats aquatic plants and usually consumes 15–20% of its body weight daily. Its preferred habitats are estuaries, saltwater bays, canals, and coastal waters. Females reach breeding maturity at five to six years and males at nine to ten years. A single calf is born every two to five years, and the calf usually stays with the mother for two years.

Range: Occasionally wanders north to Virginia and the Carolinas, sometimes even to New England. Most frequent in warm waters of southern Florida and Gulf of Mexico.

Size: Average length 10 ft. (3.5 m) and weight 1,000 lbs. (500 kg). Large individuals may reach 13 ft. (4 m) and weigh as much as 3,000 lbs (1,500 kg.).

Underwater photos of manatees are usually taken in clear freshwater environments like Florida's Crystal River Preserve. In marine coastal areas, however, the water can be almost opaque with sediment and algae, making manatees much more difficult to spot, particularly from the low vantage point of a small boat. Often you see only the nose or tail blade of a manatee lolling at the surface. This explains why boat strikes are such a common problem for manatees.

Head and open nostrils at the surface

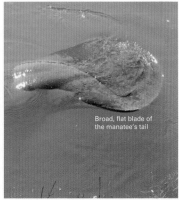

Broad, flat blade of the manatee's tail

P. LYNCH

10–13 ft. (3.5–4.5 m)

WEST INDIAN MANATEE

Average length
10 ft.

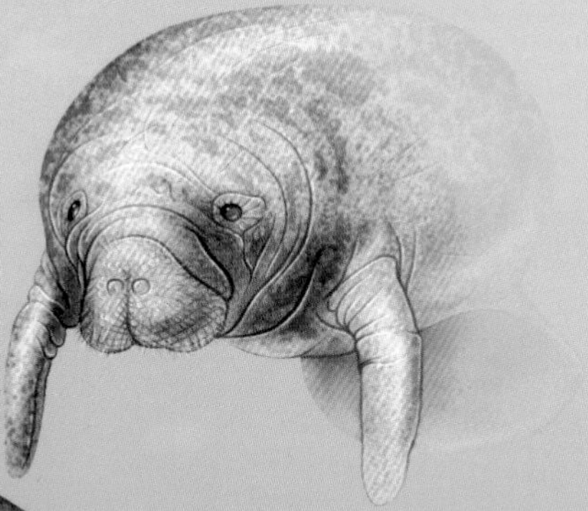

Front view, showing the very
small eyes, and the dense
whiskers of the snout

Adult and juvenile
manatee

P. LYNCH

INDEX

Page numbers in italics indicate the primary reference page for the species

Agujon, 96, *98*

Alabama River, xvi

Albacore, *162*

alula, xxviii

algae, marine, 4–7

    Common Disc Algae, *4*

    Flat-Top Bristle Brush, *4*

    Green Fleece, *6*

    Green Horsetails (Pipecleaner Algae), *4*

    Hooked Weed, *6*

    Large Sea Grapes, 6

    Mermaid's Fan, 4

    Mermaid's Wineglass, 4

    Petticoat Algae, *6*

    Pinecone Algae, 4

    Red Weed, *6*

    Sea Lettuce, *6*

Alewife, *92*

Alligator, American, *188*

Amberjack, Greater, *116*

Amberjack, Lesser, *116*

angelfish, 150–151

    Blue Angelfish, *150*

    French Angelfish, *150*

    Gray Angelfish, *150*

    Queen Angelfish, *150*

    Rock Beauty, *150*

Angel Shark, Atlantic, *80*

Angelwing, *49*

Anhinga, *216*

Antillean Current, xii

Argonaut, Greater (Paper Nautilus), *50*

Arrow Worm, *46*

Atchafalaya Bay, xvi

Atchafalaya River, xvi

Atlantic Coast, xiv

Atlantic Ocean, xii–xiii, xx

Auger, Atlantic, *48*

auriculars, xxviii

Avocet, American, *266*

Bahia Honda State Park, FL, 8

Balao, 97, *99*

Ballyhoo, 97, *99*

Bar B Ranch Preserve, Winterhaven, FL, xx

Barracuda, Great, *108*

barrier islands, xiv–xv

Bass, Black Sea, *130*

Bass, Harlequin, *154*

Bass, Striped, *108*

basslets, 154–155

    Candy Basslet, *154*

    Fairy Basslet (Royal Gramma), *154*

    Harlequin Bass, *154*

    Peppermint Basslet, *154*

bay ducks, 242–247

Bay Scallop, Atlantic, *48*

Bay Whiff (Dab), *168*

beach and dune plants, 26–29

Beach Croton, *26*

beaches, semitropical, 20–21

Beach Grass, American, *28*

beach plants, semitropical, 20–21

beaked whales, 350–353

    Blainville's Beaked Whale, *350*

    Cuvier's Beaked Whale, *352*

    True's Beaked Whale, *350*

    Sowerby's Beaked Whale, *350*

Beroe's Comb Jelly, *36*

Bigeye, *144*

billfish, 164–165

"blackfish" (toothed whales), 342–343

Black Needlerush, 18, *20*

Blood Ark, *48*

blowhole, xxix

Bluefish, *108*

Bluehead, *154*

Bonefish, *88*

Bonefish, Shafted (Longfin), *88*

Bonito, Atlantic, *160*

boobies, 210–213

    Brown Booby, *210*

    Masked Booby, *210*

    Red-Footed Booby, *212*

INDEX

brain corals, 40–41
    Boulder Brain Coral, *40*
    Grooved Brain Coral, *40*
    Symmetrical Brain Coral, *40*
branching corals, 40–41
Brant, *234*
Broomsedge, *16*
Bufflehead, *244*
Burrfish, Spotted, *178*
Butterfly, Monarch, xii
butterflyfish, 146–147
    Banded Butterflyfish, *146*
    Bank Butterflyfish, *146*
    Foureye Butterflyfish, *146*
    Longsnout Butterflyfish, *146*
    Reef Butterflyfish, *146*
    Spotfin Butterflyfish, *146*
Butterfly Ray, Smooth, *84*
Butterfly Ray, Spiny, *84*
Buttonwood, *12*

Cabbage Palmetto, *16*
Cactus, Prickly Pear, 24–25
Cannonball Jellyfish, *34*
Cape Canaveral, FL, xv
Cape Canaveral National Seashore, FL, 254
Cape Fear, NC, xv
Cape Hatteras, NC, xv, xviii, 28
Cape Lookout, NC, xv
Cape San Blas, FL, xv
Caribbean current, *xiii*
Caribbean Sea, xi
Catfish, Gafftopsail, *142*
Catfish, Hardhead, *142*
Catshark, Chain (Chain Dogfish), *78*
caudal fin (tail fin), *xxvi*
caudal keel, xxvi
caudal peduncle, xxvi
cephalopods, 50–51
Cero, *158*
Chromis, Blue, *144*
Chromis, Brown, *144*
Chub, Bermuda, *148*
Chub, Yellow, *148*

Clam, Surf, *48*
climate change, xviii
coastal hawks, 252–253
Cobia, *108*
Cockle, Giant Atlantic, *49*
cod, 94–95
Cod, Atlantic, *94*
comb jellies, 36–37
    Beroe's Comb Jelly, *36*
    Leidy's Comb Jelly, *36*
    Northern Comb Jelly, *36*
    Sea Gooseberry, *36*
    Venus Girdle, *36*
Continental shelf, xiv
coral reefs, 38–39
corals, 38–45
    Bladed Fire Coral, *42*
    Boulder Brain Coral, *40*
    Elkhorn Coral, *40*
    Grooved Brain Coral, *40*
    Maze Coral, *40*
    Mountainous Star Coral, *42*
    Pillar Coral, *42*
    Smooth Star Coral, *42*
    soft corals, 44–45
    Staghorn Coral, *40*
    Symmetrical Brain Coral, *40*
Cordgrass, Saltmeadow, *20*
Cordgrass, Smooth, *20*
Cormorant, Double-Crested, *214*, 216
Cormorant, Great, *214*
Corolla, NC, 248, 276
Coronetfish, Bluespotted, 96, *98–99*
coverts (feathers), xviii
Cowfish, Honeycomb, *172*
Cowfish, Scrawled, *172*
crabs, 56–61
    Asian Shore Crab, *60*
    Atlantic Rock Crab, *58*
    Blue Crab, *56*
    Common Spider Crab, *60*
    Flat-Clawed Hermit Crab, *60*
    Ghost Crab, *58*
    Horseshoe Crab, xiii, *60*

Jonah Crab, *58*

Lady Crab, *56*

Sand Fiddler Crab, *58*

Stone Crab, *56*

Striped Hermit Crab, *60*

Croaker, Atlantic, *132*

Crocodile, American, *188*

Cross-Barred Venus, *48*

ctenophores (comb jellies), 36–37

Cunner, *130*

Curlew, Long-Billed, *272*

Currituck, NC, 20

Dab (Bay Whiff), *168*

damselfish, 144–145

    Bicolor Damselfish, *144*

    Cocoa Damselfish, *144*

    Yellowtail Damselfish, *144*

Deepwater Horizon, xvii, xix

DeSoto Canyon, xxv

Devilfish (Atlantic Manta), *86*

Disk Dosina, *48*

Doctorfish, *148*

dogfish, 78–79

    Black Dogfish, *78*

    Chain Catshark (Chain Dogfish), *78*

    Cuban Dogfish, *78*

    Smooth Dogfish, *78*

    Spiny Dogfish, *78*

    topography of, xxvi

Dogwood, *19*

dolphin (fish), 118

    Dolphin (Dorado), *118*

    Pompano Dolphin, *118*

dolphins (marine mammals), 354–356

    Atlantic Spotted Dolphin, *356*, 365

    Atlantic White-Sided Dolphin, *354*, 364

    Bottlenose Dolphin (coastal form), *362*, 364

    Bottlenose Dolphin (offshore form), *362*, 364

    Clymene Dolphin, *358*

    Common Dolphin, *360*, 364

    Pantropical Spotted Dolphin, *356*, 365

    in profile, 364–365

    Rough-Toothed Dolphin, *360*, 365

Spinner Dolphin, *358*, 365

Striped Dolphin, *358*, 365

dorsal fin, xxvi, xxvii

Dowitcher, Long-Billed, *274*

Dowitcher, Short-Billed, *274*

drums, 132–135

    Atlantic Croaker, *132*

    Black Drum, *132*

    High-Hat, *134*

    Jackknife Fish, *134*

    Red Drum, *132*

    Spot, *132*

    Spotted Drum, *134*

Dry Tortugas National Park, FL, 192, 308, 312

ducks, 240–247

    bay, 242–245

    Black Scoter, *240*, 247

    Bufflehead, *244*, 246

    Common Eider, *240*, 247

    Common Goldeneye, *242*, 246

    Greater Scaup, *242*, 247

    Lesser Scaup, *242*, 246

    Long-Tailed Duck, *244*, 246

    Red-Breasted Merganser, *244*, 246

    sea, in flight, 246–247

    Surf Scoter, *240*, 247

    White-Winged Scoter, *240*, 247

dune and beach plants, 26–27

Dunlin, *274*

Durgon, Black, *176*

Eagle, Bald, *250*, 256

Eagle Ray, Spotted, *86*

Eagle Ray, Southern, *84*

Eelgrass, *8*, 236

eels, 90–91

    American Eel, *90*

    Chain Moray, *90*

    Conger Eel, *90*

    Green Moray, *90*

    Spotted Moray, *90*

egrets, 222–225

    Cattle Egret, xx, *224*

    Great Egret, *224*, 226

INDEX

egrets (*continued*)
    Reddish Egret, *222*
    Snowy Egret, *224*
Eider, Common, *240*

Fiddler Crab, Sand, *58*
filefish, 174–175
    Orange Filefish, *174*
    Orangespotted Filefish, *174*
    Planehead Filefish, *174*
    Scrawled Filefish, 38, *174*
    Unicorn Filefish, *174*
    Whitespotted Filefish, *174*
Fire Coral, Bladed, *42*
fish, topography, xxvii
Flamingo, Greater, *228*
Flat-Top Bristle Brush, *4*
flippers, xxix
Florida, xiv, xv, xvi. *See also specific locations*
Florida current, xii, xiii
Florida Keys, xii, 10, 14
Florida Straits, xii
flounders, 166–169
    Atlantic Halibut, *168*
    Bay Whiff (Dab), 168
    Gulf Flounder, *166*
    Gulf Stream Flounder, *166*
    lefteye, 166–167
    Southern Flounder, *166*
    Summer Flounder, *166*
    Windowpane, *166*
    Winter Flounder, *168*
flyingfish, 100–101
    Atlantic Flyingfish, *100*
    Bandwing Flyingfish, *100*
    Margined Flyingfish, *100*
    Oceanic Two-Wing Flyingfish, *100*
    Spotfin Flyingfish, *100*
foredunes and foredune community, xiv–xv
Franklin, Benjamin, xii
Frigatebird, Magnificent, *192*, 256
Frogfish, Longlure, *140*
Fulmar, Northern, *194*

Gannet, Northern, *208*, 211, 213
geese, 236–239
    "Blue goose," 238
    Brant, 236
    Canada Goose, 236
    Greater White-Fronted Goose, 238
    Snow Goose, 238
Georgia, xiv
gill cover, xxvii
gill slits, xxvi
glassworts, 28–29
Glaucus, Blue (sea snail), 50
Goatfish, Yellow, *134*
Goatfish, Spotted, *134*
Goldeneye, Common, 242
Golden-Plover, American, *260*
Goosefish, 142
gorgonians (soft corals), 44–45
Grampus (Risso's Dolphin), *344*
Grand Banks, xii
grass, marine, 8–9
    Eelgrass, *8*
    Manatee Grass, *8*
    Turtle Grass, *8*
Grebe, Horned, *190*
Grebe, Pied-Billed, *190*
Green Horsetails, *4*
groupers, 102–105
    Black Grouper, *106*
    Gag Grouper, *106*
    Goliath Grouper, *106*
    Nassau Grouper, *102*
    Red Grouper, *102*
    Red Hind, *102*
    Rock Hind, *104*
    Sand Perch, *104*
    Scamp, *104*
    Snowy Grouper, *104*
    Speckled Hind, *102*
    Warsaw Grouper, *106*
    Yellowedge Grouper, *104*
grunts, 124–125
    Black Margate, *124*
    Bluestriped Grunt, *124*

French Grunt, *124*
Margate, *124*
Porkfish, *124*
Tomtate, *124*
White Grunt, *124*
Guitarfish, Atlantic, *80*
Gulf Coast, xvi
Gulf Gyre, xii, xvi
Gulf Loop Current, xiii
Gulf of Mexico, xi, xvi, xxii–xxiii, xxiv–xxv
Gulf Stream, xi–xiii
Gulfweed, *2*
gulls, 284–301
    adult, in flight, 298–299
    Black-Legged Kittiwake, *296*, 299
    Bonaparte's Gull, *294*, 298, 300
    Franklin's Gull, 290, *292*, 298, 300
    Glaucous Gull, *298*
    Great Black-Backed Gull, 256, *284*, 299, 301
    Herring Gull, *286*, 299, 301
    Iceland Gull, *298*
    immature, in flight, 300–301
    Laughing Gull, 290, *292*, 298, 300
    Lesser Black-Backed Gull, *299*
    Little Gull, *298*, 300
    Ring-Billed Gull, *288*, 299, 301
    Sabine's Gull, *296*, 299
    topography of, xxviii
Gyre, Gulf, xii, xvi
Gyre, North Atlantic, xii

habitat destruction, xix
Haddock, *94*
Hake, Silver, *94*
Hake, White, *94*
Halfbeak, American, *97*, 99
Halibut, Atlantic, *168*
hammerheads, 76–77
    Great Hammerhead, *76*
    Scalloped Hammerhead, *76*
    Smooth Hammerhead, *76*
hawks, 252–253
    Broad-Winged Hawk, *252*, 256
    Red-Shouldered Hawk, *252*, 256

Red-Tailed, Hawk, *252*, 256
Hermit Crab, Flat-Clawed, *60*
Hermit Crab, Striped, *60*
herons, 220–227
    Cattle Egret, xx, *224*
    Great Blue Heron, *226*
    Great Egret, *224*, 226
    "Great White Heron," *226*
    Green Heron, *220*
    Little Blue Heron, *222*
    Reddish Egret, *222*
    Snowy Egret, *224*
    Tricolored Heron, *220*
    "Würdemann's Heron," *226*
Herring, Atlantic, *92*
Herring, Blueback, *92*
High-Hat, *134*
Hind, Red, *102*
Hind, Rock, *104*
Hogchoker, *170*
Hogfish, *152*
Holly, Yaupon, xiv, 19
Houndfish, *97*, 98–99
hurricanes, xvi–xvii
    Galveston (1900), xvi
    Hurricane Camille, xvi
    Hurricane Katrina, xvi
    Hurricane Mitch, xvii

ibises, 232–233
    Glossy Ibis, *232*
    Scarlet Ibis, *233*
    White-Faced Ibis, *232*
    White Ibis, 206, *232*
IUCN Red List, xviii, xx
Ivy, Poison, *28*

Jackknife Fish, *134*
jacks, 112–117
    Almaco Jack, *114*
    Banded Rudderfish, *112*
    Bar Jack, *112*
    Blue Runner, *112*
    Bluntnose Jack, *116*

INDEX

jacks (*continued*)
    Crevalle Jack, *114*
    Greater Amberjack, *116*
    Horse-Eye Jack, *114*
    Leatherjack (Leatherjacket), *112*
    Lesser Amberjack, *116*
    Pilotfish, *112*
    Rainbow Runner, *116*
    Yellow Jack, *114*
jaegers, 282–283
    Long-Tailed Jaeger, *282*
    Parasitic Jaeger, *282*
    Pomarine Jaeger, *282*
Janthina, Common, *49*
jellyfish, 30–35
    By-the-Wind Sailor, *30*
    Cannonball Jellyfish, *34*
    Lion's Mane Jellyfish, *32*
    Mangrove Upside-Down Jellyfish, *34*
    Moon Jelly, *32*
    Mushroom Jellyfish, *32*
    Portuguese Man-of-War, *30*
    Sea Nettle, *30*
    Sea Wasp, *30*
    Upside-Down Jellyfish, *34*
John Pennekamp Coral Reef State Park, FL, 10, 38–39

Kemp's Ridley, *184*
Kestrel, American, *252*
Key Largo, FL, *10*
Killdeer, *264*
Killer Whale, False, *340*, 342
Killer Whale (Orca), 342, *346*
kingfish, 136–137
    Gulf Kingfish, *136*
    Northern Kingfish, *136*
    Southern Kingfish, *136*
Kittiwake, Black-Legged, *296*
Knot, Red, *274*

Labrador Current, xiii
Lace Murex, *49*
Ladyfish, *88*

Lake Mattamuskeet National Wildlife Refuge, NC, 224
lateral lamina, xxix
lateral line, xxvi, xxvii
Leatherjack (Leatherjacket), *112*
Leidy's Comb Jelly (Sea Walnut), *36*
León, Ponce de, xii
Lettered Olive Shell, *49*
Lion's Mane Jellyfish, *32*
lobsters, 54–55
    Caribbean Spiny Lobster, *54*
    Northern Lobster, *54*
    Spanish Lobster, *54*
longshore currents, xiv
longshore drift, xiv
Lookdown, *110*
Loon, Common, *190*
Loon, Red-Throated, *190*
lores, xxviii
Louvar, *118*

mackerel, 158–161
    Atlantic Mackerel, *158*
    Bullet Mackerel, *160*
    Cero, *158*
    Frigate Mackerel, *160*
    King Mackerel, *158*
    Spanish Mackerel, *158*
    Wahoo, *158*
Macondo well, xvii
Mako, Longfin, *66*
Mako, Shortfin, *66*
Manatee, West Indian, *370*
Manatee Grass, *8*
mangal, *10*
mangroves, 10–17
    Black Mangrove, 10, *12*, 14–15
    communities, 10
    Red Mangrove, 10, *12*
    and salt marshes, 14–15, 16–17
    White Mangrove, 10, *12*, 14–15, 220
Manta, Atlantic (Devilfish), *86*
Margate, *124*
Margate, Black, *124*

marine algae, 4–7
marine grasses, 8–9
maritime forests, xv
Marlin, Blue, *164*
Marlin, White, *164*
melon, in cetaceans, xxix
Menhaden, Atlantic (Mossbunker), *92*
Merganser, Red-Breasted, *244*
Mermaid's Fan, 4, *8*
Mermaid's Wineglass, *4*
Merritt Island National Wildlife Refuge, FL, 14, 220, 228
Mexico Basin, xxiv
migration, xii
Mississippi Fan, xxv
Mississippi River, xvi
Mobile Bay, xvi
Mola, Sharptail, *180*
"Monkfish" (Goosefish), *142*
Moon Jellyfish, *32*
Moon Snail, Atlantic, *48*
moray eels, 90–91
 Chain Moray, *60*
 Green Moray, *60*
 Spotted Moray, *60*
Mossbunker (Atlantic Menhaden), *92*
Mullet, Striped, *136*
Mullet, White, *136*
Murex, Lace, *49*

Nag's Head Woods Ecological Preserve, NC, xi, 18–19
Needlefish, Atlantic, *97*, 99
Needlefish, Flat, *96*, 98
Night-Heron, Black-Crowned, *218*
Night-Heron, Yellow-Crowned, *218*
Noddy, Black, *312*
Noddy, Brown, *312*
North Atlantic Gyre, xii
North Carolina, xi, xv. *See also specific locations*
North Equatorial Current, xii
Northern Comb Jelly, *36*
Northern Fulmar, *194*

Oak, Laurel, *19*
Oak, Live, *20*
Oats, Sea, 22–23, *24*, 26–27, 28
ocean sunfish, 180–181
Oikopleura, *46*
oiled birds, xviii
oil platforms, xx
oil spills, xvii
Opah, *118*
Orca (Killer Whale), 342, *346*
Osprey, *248*, 256
overfishing, xix
Oyster, Eastern, *49*
Oystercatcher, American, xviii, *266*

Padre Island, TX, xv
palmettos, 14, 16–17, 24–25
 Cabbage Palmetto, 14, *16*, 24–25
 Saw Palmetto, 22–23, *24*
Palometa, *110*
Paper Nautilus (Argonaut), *50*
parrotfish, 156–157
 Blue Parrotfish, *156*
 Queen Parrotfish, *156*
 Rainbow Parrotfish, *156*
 Stoplight Parrotfish, 39, *156*
pectoral fin, xxvi, xxvii
pelagic invertebrates, 46–47
pelagic rays, 46–46
pelicans, xxviii, 204–207
 Brown Pelican, *204*, 256
 topgraphy of, xxviii
 White Pelican, *206*, 256
Perch, Sand, *104*
Perch, White, *138*
Permit, *110*
Petrel, Black-Capped, *196*
Petticoat Algae, *6*
phalaropes, 258–259
 Red-Necked Phalarope, *258*
 Red Phalarope, *258*
 Wilson's Phalarope, *258*
Pilotfish, 70–71, *112*
Pilot Whale, Long-Finned, *336*, 342

INDEX

Pilot Whale, Short-Finned, *338,* 342

Pinecone Algae, *4*

pines, 18, 20–21, 24–25

    Loblolly Pine, 18, *20*

    Longleaf Pine, 18

    Slash Pine, *24*

Pinfish, *126*

Pipecleaner Algae, *4*

pipefish, 96–99

    Dusky Pipefish, *96,* 98

    Gulf Pipefish, *96,* 98

    Sargassum Pipefish, *96,* 98

plankton, 46–47

Plankton Worm, *46*

Playalinda Beach, FL, 254

plovers, 260–265

    American Golden-Plover, *260*

    Black-Bellied Plover, *60*

    Killdeer, *264*

    Piping Plover, xix, *262*

    Semipalmated Plover, *264*

    Snowy Plover, *262*

    Wilson's Plover, *262*

pneumatophores, mangrove, 12

Poison Ivy, *28*

Pollock, *94*

pompanos, 110–111

    African Pompano, *110*

    Florida Pompano, *110*

    Lookdown, *110*

    Palometa, *110*

    Permit, *110*

Porbeagle, *66*

Porcupinefish, *178*

porgies, 126–129

    Grass Porgy, *128*

    Jolthead Porgy, *126*

    Knobbed Porgy, *126*

    Littlehead Porgy, *128*

    Pinfish, *126*

    Red Porgy, *126*

    Saucereye Porgy, *128*

    Sea Bream, *128*

    Sheepshead Porgy, *126*

    Silver Porgy, *128*

    Spottail Porgy, *126*

    Whitebone Porgy, *128*

Porkfish, *124*

Porpoise, Harbor, *354*

porpoises. *See* dolphins (marine mammals);

    Porpoise, Harbor

Portuguese Man-of-War, *30*

precaudal pit, xxvi

preopercle, xxvii

primary feathers, xxviii

Puddingwife, *156*

puffers, 178–179

    Northern Puffer, *178*

    Oceanic Puffer, *178*

    Smooth Puffer, *178*

Quahog, Southern, *49*

Railroad Vine, *26*

Rainbow Runner, *116*

range maps, about, xx

rays, 84–87

    Atlantic Manta (Devilfish), *86*

    Bullnose Ray, *84*

    Cownose Ray, *84*

    Devil Ray, *86*

    skates, 80–81

    Smooth Butterfly Ray, *84*

    Southern Eagle Ray, *84*

    Spiny Butterfly Ray, *84*

    Spotted Eagle Ray, *86*

    stingrays, 82–83

    topography of, xxvi

Red-Cedar, Eastern, *19*

Red Knot, xiii

Red List, IUCN, xviii, xx

red tide, xviii

Red Weed, *6*

reef balls, xix

reef fish, small, 144–145

Reef Shark, Caribbean, *72*

Remora, *70*

Right Whale, Northern, *332*

Risso's Dolphin (Grampus), *344*

Roanoke Sound, NC, xi

Rock Beauty, *150*

Rock Crab, Atlantic, *58*

Royal Gramma (Fairy Basslet), *154*

Rudderfish, Banded, *112*

Runner, Blue, *112*

Runner, Rainbow, *116*

Sailfish, *164*

Salp, *46*

salt marshes, 14–15, 18–19

salt-spray zone, xv

Sandbur, *26*

Sanderling, *276*

sandpipers, 270–271, 276–279

 Least Sandpiper, *278*

 Purple Sandpiper, *270*

 Semipalmated Sandpiper, *278*

 Spotted Sandpiper, *270*

 Western Sandpiper, *278*

 White-Rumped Sandpiper, *276*

Sargasso Sea, *2*

Sargassumfish, 2, *140*

Sargassum Triggerfish, 2, *176*

scallops, 48, 49

 Atlantic Bay Scallop, *48*

 Calico Scallop, *49*

Scamp, *104*

Scaup, Greater, *242*

Scaup, Lesser, *242*

Schoolmaster, *120*

Scorpionfish, Spotted, *140*

scoters, 240–241

 Black Scoter, *240*

 Surf Scoter, *240*

 White-Winged Scoter, *240*

Scup, *130*

Sea Bass, Black, *130*

Seabeach Orache, *27*

Sea Bream, *128*

Sea Butterfly, Naked, *46*

Sea Butterfly, Shelled, *46*

Sea Fan, Common, 38–39, 44

Sea Fingers, Corky, 39, *44*

Sea Gooseberry, *36*

Sea Grape, *20*

Sea Grapes, Large, *6*

seahorses, 140–141

 Dwarf Seahorse, *140*

 Lined Seahorse, *140*

 Longsnout Seahorse, *140*

Sea Lettuce, *6*, 236

seals, 366–369

 Gray Seal, xviii, *368*

 Harbor Seal, *368*

Sea Oats, 22–23, *24*, 26–27, 28

Sea Plume, Rough, *39*

Sea Rod, Bent, *44*

seatrout, 138–139

 Sand Seatrout, *138*

 Silver Seatrout, *138*

 Spotted Seatrout, *138*

sea turtles, 9, 182–187

 Green Turtle, 182

 Hawksbill Turtle, 182

 Kemp's Ridley, *9*, 184

 Leatherback Sea Turtle, 186

 Loggerhead Sea Turtle, 184

Sea Walnut (Leidy's Comb Jelly), *36*

secondary feathers, xxviii

Sergeant Major, *144*

Shad, American, *92*

Shad, Hickory, *92*

sharks, 66–79, 351

 Atlantic Sharpnose Shark, *72*

 Basking Shark, *64*

 Bigeye Thresher, *76*

 Blacktip Shark, *68*

 Blue Shark, *66*

 Bonnethead Shark, *76*

 Bull Shark, *74*

 Caribbean Reef Shark, *72*

 Cookiecutter Shark, *351*

 Dusky Shark, *74*

 Great Hammerhead Shark, *76*

 inshore, 74–75

 Lemon Shark, *72*

INDEX

sharks (*continued*)

Longfin Mako Shark, *66*

Night Shark, *72*

Nurse Shark, *74*

Oceanic Whitetip Shark, *68*

Porbeagle, *66*

Sandbar Shark (Brown Shark), *74*

Sand Tiger (Sand Shark), *74*

Scalloped Hammerhead Shark, *76*

Shortfin Mako Shark, *66*

Silky ("Sickle") Shark, *68*

Sixgill Shark, *78*

Smooth Hammerhead Shark, *76*

Spinner Shark, *68*

Thresher Shark, *76*

Tiger Shark, *68*

topography of, xxvi

Whale Shark, *62*

White Shark, *66*

Sharksucker, *70*

Sharksucker, Whitefin, *70*

Sharpnose Shark, Atlantic, *72*

shearwaters, 196–199

Audubon's Shearwater, *198*

Cory's Shearwater, *196*

Greater Shearwater, *196*

Manx Shearwater, *198*

Sooty Shearwater, *196*

shells, common beach, 48–49

Shore Crab, Asian, *60*

shrimp, 2, 52–53

Brown Shrimp, *52*

Pink Shrimp, *52*

Slender Sargassum Shrimp, 2, *52*

White Shrimp, *52*

skates, 80–81

Barndoor Skate, *80*

Clearnose Skate, *80*

Winter Skate, *80*

Skimmer, Black, xviii, *314*, 317, 319

Skua, Great, *280*

Skua, South Polar, *280*

Slash Pine, *16*

Slippery Dick, *154*

soft corals (gorgonians), 44–45

snappers, 120–123

Blackfin Snapper, *122*

Cubera Snapper, *120*

Dog Snapper, *122*

Gray Snapper, *120*

Lane Snapper, *122*

Mutton Snapper, *122*

Queen Snapper, *120*

Schoolmaster, *120*

Silk Snapper, *122*

Vermillion Snapper, *120*

Yellowtail Snapper, *120*

Soldierfish, Slackbar, *144*

Soldierfish, Cardinal, *144*

soles, 170–171

Fringed Sole, *170*

Hogchoker, *170*

Naked Sole, *170*

Scrawled Sole, *170*

South Carolina, xiv. *See also specific locations*

South Equatorial Current, xi

Spadefish, Atlantic, *152*

Spearfish, Longbill, *164*

Sperm Whale, Dwarf, *348*

Sperm Whale, Pygmy, *348*

Spider Crab, Common, *60*

spiny-rayed fish, topography, xxvii

spiracle, xxvi

splashguard, in whales, xxix

Sponge, Branching Tube, *44*

Spoonbill, Roseate, *230*

Spot, *132*

Spotted Dolphin, Atlantic, *356*, 365

Spotted Dolphin, Pantropical, *356*, 365

squid, 50–51

Atlantic Brief Squid, *50*

Caribbean Reef Squid, *50*

Longfin Inshore Squid, *50*

Squirrelfish, *144*

Star Coral, Mountainous, *42*

Star Coral, Smooth, *42*

Stilt, Black-Necked, *266*

INDEX

stingrays, 82–83
  Atlantic Stingray, *82*
  Bluntnose Stingray, *82*
  Pelagic Stingray, *82*
  Roughtail Stingray, *82*
  Southern Stingray, *82*
Stork, Wood, 206, *228*
storm-petrels, 200–203
  Band-Rumped Storm-Petrel, *202*
  Leach's Storm-Petrel, *200*
  White-Faced Storm-Petrel, *202*
  Wilson's Storm-Petrel, *200*
Straits of Florida, xxiii
Sunfish, Ocean, *180*
Surgeon, Ocean, *148*
Surgeonfish, *148*
Swan, Mute, *234*
Swan, Tundra, *234*
Swordfish, *164*

Tang, Blue, *148*
Tarpon, *88*
Tautog, *130*
terns, 302–319
  adult, in flight, 316–317
  Black Noddy, *312*, 317, 318
  Black Tern, *314*, 317, 318
  Bridled Tern, *310*, 317, 318
  Brown Noddy, *312*, 317, 319
  Caspian Tern, *308*, 316, 319
  Common tern, *302*, 316, 318
  Forster's Tern, *302*, 316, 318
  Gull-Billed Tern, *306*, 316, 318
  Least Tern, *304*, 316, 318
  Roseate Tern, xix, *304*, 317, 318
  Royal Tern, *308*, 316, 319
  Sandwich Tern, *306*, 316, 319
  Sooty Tern, *310*, 317, 319
  standing, size comparison, 318–319
Texas, xvi. *See also specific locations*
Thistle, Russian, *26*
Thresher, Bigeye, *76*
Tiger, Sand (Sand Shark), *74*
Tilefish, *152*

Tilefish, Sand, *152*
Toadfish, Gulf, *142*
Toadfish, Oyster, *142*
Tobaccofish, *154*
Tomtate, *124*
topography, xxvi–xxix
  birds, xxviii
  dolphins, xxix
  fish, xxvii
  sea turtles, xxix
  sharks and rays, xxvi
  whales, xxix
Torpedo, Atlantic, *80*
triggerfish, 176–177
  Black Durgon, *176*
  Gray Triggerfish, *176*
  Ocean Triggerfish, *176*
  Queen Triggerfish, *176*
  Sargassum Triggerfish, *176*
Tripletail, *106*
Tropicbird, White-Tailed, *192*
Trumpetfish, *97*, 98–99
trunkfish, 172–173
  Smooth Trunkfish, *172*
  Spotted Trunkfish, *172*
  Trunkfish, *172*
tuna, 160–163
  Albacore, *162*
  Atlantic Bonito, *160*
  Bigeye Tuna, *162*
  Blackfin Tuna, *160*
  Bluefin Tuna, xix, *162*
  Bullet Mackerel, *160*
  Frigate Mackerel, *160*
  Little Tunny, *160*
  Skipjack Tuna, *160*
  Yellowfin Tuna, *162*
Tunny, Little, *160*
Turtle Grass, *8*
turtles. *See* sea turtles
Turnstone, Ruddy, *270*
typical dolphin, topograpy, xxix
typical fish, topography, xxvii
typical shark, topography, xxvii

INDEX

Venus Girdle, *36*

vertebral lamina, xxix

Virginia Creeper, 28

Vulture, Black, *254*, 256

Vulture, Turkey, *254*, 256

Wahoo, *158*

Wax Myrtle, 28

Weakfish, *138*

West Florida Shelf, xxiii

West Indian faunal province, xvi

whales, xiv, 320–353

    "blackfish" (toothed whales), 342–343

    Blainville's Beaked Whale, *350*

    Blue Whale, *324*

    Bryde's Whale, *328*

    Cuvier's Beaked Whale, *352*

    Dwarf Sperm Whale, *348*

    False Killer Whale, *340*

    Fin Whale, xiv, *320*

    Humpback Whale, xiv, *330*

    Killer Whale (Orca), *346*

    Long-Finned Pilot Whale, *336*

    Minke Whale, *322*

    Northern Right Whale, xiv, *332*

    Pygmy Sperm Whale, *348*

    Sei Whale, *326*

    Short-Finned Pilot Whale, *336*

    Sowerby's Beaked Whale, *350*

    Sperm Whale, *334*

    topography of, xxviii

    True's Beaked Whale, *350*

whelks, 48–49

    Channeled Whelk, *48*

    Knobbed Whelk, *48*

    Lightning Whelk, *49*

    Pear Whelk, *48*

Whimbrel, *272*

Whitetip Shark, Oceanic, *68*

Willet, *268*

Windowpane, *166*

wrasses, 154–155, 156

    Creole Wrasse, *154*

    Puddingwife, *156*

    Yellowcheek Wrasse, *154*

Yellowlegs, Greater, *268*

Yellowlegs, Lesser, *268*

Yucatán Channel, xvi, xxiii, xxv

Yucatán Current, xvi